ADVANCES IN CHEMICAL PHYSICS

VOLUME 137

EDITORIAL BOARD

ADVANCES IN CHEMICAL PHYSICS

VOLUME 137

Series Editor

STUART A. RICE

Department of Chemistry
and
The James Franck Institute
The University of Chicago
Chicago, Illinois

WILEY-INTERSCIENCE

A JOHN WILEY & SONS, INC. PUBLICATION

Published by John Wiley & Sons, Inc., Hoboken, New Jersey
Published simultaneously in Canada

For general information on our other products and services or for technical support, please contact our Customer Care Department within the United States at (800) 762-2974, outside the United States at (317) 572-3993 or fax (317) 572-4002.

Wiley also publishes its books in a variety of electronic formats. Some content that appears in print may not be available in electronic formats. For more information about Wiley products, visit our web site at www.wiley.com.

Library of Congress Catalog Number: 58-9935

ISBN 978-0-471-43573-0

Printed in the United States of America

10 9 8 7 6 5 4 3 2 1

CONTRIBUTORS TO VOLUME 137

SAVO BRATOS, Laboratoire de Physique Théorique des Liquides, Université Pierre et Marie Curie, Case Courrier 121, 4 Place Jussieu, Paris Cedex 75252, France

JACK F. DOUGLAS, Polymers Division, National Institute of Standards and Technology, Gaithersburg, MD 20899 USA

JACEK DUDOWICZ, The James Franck Institute and the Department of Chemistry, The University of Chicago, Chicago, IL 60637 USA

KARL F. FREED, The James Franck Institute and the Department of Chemistry, The University of Chicago, Chicago, IL 60637 USA

FELIX RITORT, Department de Fisica Fonamental, Faculty of Physics, Universitat de Barcelona, Diagonal 647, 08028 Barcelona, Spain

MICHAEL WULFF, European Synchrotron Radiation Facility, Grenoble Cedex 38043, BP 220, France

INTRODUCTION

Few of us can any longer keep up with the flood of scientific literature, even in specialized subfields. Any attempt to do more and be broadly educated with respect to a large domain of science has the appearance of tilting at windmills. Yet the synthesis of ideas drawn from different subjects into new, powerful, general concepts is as valuable as ever, and the desire to remain educated persists in all scientists. This series, *Advances in Chemical Physics*, is devoted to helping the reader obtain general information about a wide variety of topics in chemical physics, a field that we interpret very broadly. Our intent is to have experts present comprehensive analyses of subjects of interest and to encourage the expression of individual points of view. We hope that this approach to the presentation of an overview of a subject will both stimulate new research and serve as a personalized learning text for beginners in a field.

STUART A. RICE

CONTENTS

TIME-RESOLVED X-RAY DIFFRACTION FROM LIQUIDS

SAVO BRATOS

Laboratoire de Physique Théorique de la Matière Condensée, Université Pierre et Marie Curie, Case Courrier 121, 4 Place Jussieu, Paris Cedex 75252, France

MICHAEL WULFF

European Synchrotron Radiation Facility, Grenoble Cedex 38043, BP220, France

CONTENTS

Advances in Chemical Physics, Volume 137, edited by Stuart A. Rice

I. INTRODUCTION

Since the discovery of X-rays by Roentgen, X-ray diffraction has always been the major technique permitting the localization of atoms in molecules and crystals. Great scientists such as Bragg, Laue, and Debye made major contributions to its development. Atomic structures of many systems have been determined using this technique. These structures are actually known with a great accuracy, and one can hardly imagine a science without this information. The recent progress achieved using synchrotron sources of X-ray radiation are impressive.

However, systems with localized atoms represent only a first challenge. The next challenge is monitoring atomic motions in systems that vary in time. Following atomic motions during a chemical process has always been a dream of chemists. Unfortunately, these motions evolve from nanosecond to femtosecond time scales, and this problem could not have been overcome until ultrafast detection techniques were invented. Spectacular developments in laser technology, and recent progress in construction of ultrafast X-ray sources, have proved to be decisive. Two main techniques are actually available to visualize atomic motions in condensed media.

The first is time-resolved optical spectroscopy. The system is excited by an intense optical pulse, and its return to statistical equilibrium is probed by another pulse, which is also optical. Zewail and several other outstanding scientists contributed much to its development. For textbooks describing it, see Refs. 1–4. The second technique comprises time-resolved X-ray diffraction and absorption. The excitation of the system is optical as before, but the probing is done using an X-ray pulse. Long range order may be probed by diffraction, whereas short range order may be monitored by absorption. Unfortunately, general literature is still scarce in this domain [5–7].

The major advantage of time-resolved X-ray techniques, as compared to optical spectroscopy, is that their wavelength λ as well as the pulse duration τ can be chosen to fit the atomic scales. This is not the case for optical spectroscopy, where the wavelength λ exceeds interatomic distances by three orders of magnitude at least. Unfortunately, X-ray techniques also have their drawbacks. They require large scale instruments such as the synchrotron. Even much larger instruments based on free electron lasers are actually under construction. The "nonhuman" size of X-ray instrumentation is sometimes an objection against the use of this method, whereas optical spectroscopy is free of this objection.

The purpose of this chapter is to review ultrafast, time-resolved X-ray diffraction from liquids. Both experimental and theoretical problems are treated. Section II describes the principles of a time-resolved X-ray experiment and details some of its characteristics. Basic elements of the theory are discussed in Sections III–V. Finally, Section VI presents recent achievements in this domain. The related field of time-resolved X-ray spectroscopy, although very promising, is not discussed.

II. EXPERIMENT

A. Basic Principles

The system under consideration is a liquid sample, either a pure liquid or a solution. It is pumped by a laser, which promotes a fraction of the molecules into one or several excited quantum states. The energy deposited by the laser diffuses into the system in one or several steps, generating several sorts of events. If the excitation is in the optical spectral range, chemical reactions may be triggered and the sample is heated. If it is in the infrared, only heating is generally present. The return of the excited system to thermal equilibrium is then probed using a series of time-delayed X-ray pulses (Fig. 1). The resulting diffraction patterns consists of circular rings, centered on the forward beam direction. Finally, the collection of diffraction patterns obtained in this way is transformed into a collection of molecular photographs; this step is accomplished by theory. A time-resolved X-ray experiment thus permits one to "film" atomic motions.

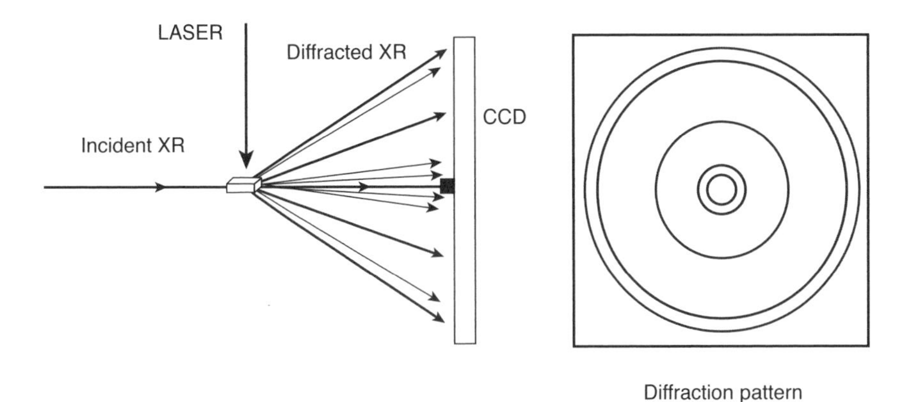

Figure 1. Time-resolved X-ray diffraction experiment (schematic). The liquid sample is excited by a laser pulse, and its temporal evolution is monitored by a time-delayed X-ray pulse. The diffracted radiation is measured by a charge-coupled detector (CCD). In practice, the laser and X-ray beams are not perpendicular to each other, but nearly parallel.

This rough picture can be sharpened by providing some additional information. The experimental setup appropriate for time-resolved diffraction comprises a pulsed synchrotron source, a chopper that selects single X-ray pulses from the synchrotron, a femtosecond laser activating the process to be studied, a capillary jet, and an integrating detector measuring the intensity of the scattered X-ray radiation. What is desired in reality is not the scattered X-ray intensity by itself, but the difference between scattered intensities in the presence or absence of the laser excitation. This difference intensity is very small and is thus particularly difficult to measure. The images must be integrated azimuthally and corrected for polarization and space-angle effects. How is this sort of experiment realized in practice? Some major points are discussed next [5–8]. For new trends in instrumentation, see Refs. 9–11.

B. Practical Realization

1. X-Ray and Optical Sources

The first and central point in this discussion concerns the pulsed X-ray source. The shortest time scales involved when chemical bonds are formed or broken are on the order of a few femtoseconds. An ideal X-ray source should thus be capable of providing pulses of this duration. Unfortunately, generating them represents a heavy technological problem. The best one can do at present is to use a pulsed synchrotron X-ray source (Fig. 2). Electrons are rapidly circulating in a storage ring at speeds close to the speed of light. X-rays are spontaneously emitted longitudinal to the orbit. This emission is amplified in so-called straight sections,

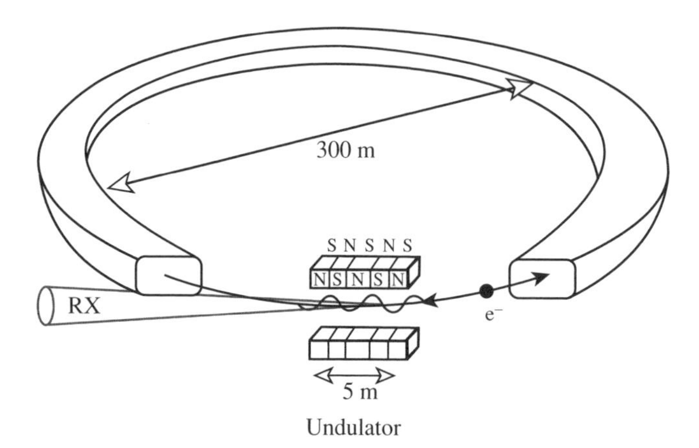

Figure 2. Synchrotron X-ray source (schematic). The electron execute circular motions in the storage ring and emit intense X-rays along the tangent of the orbit. This radiation is enhanced by undulator magnets that are often placed inside the vacuum vessel for enhanced performance. The storage ring has a number of straight sections for undulators and wigglers (not shown).

where a sinusoidal motion is imposed on the electrons by undulator magnets. The X-rays emitted from successive bends interfere and enhance the radiation. A bunched electron beam then produces an intense X-ray radiation with 100 ps X-ray pulses. Unfortunately, subpicosecond X-ray pulses cannot be generated in this way. There is a gap between what is possible at present and what is needed. Facilities based on the use of free electron lasers are under construction to bridge this gap.

Sources of optical radiation are much more conventional. One generally employs commercially available lasers, generating 100 fs pulses with pulse energy between 10 and 100 μJ. These lasers run in phase with the chopper. The power density of the optical beam on the sample is typically on the order of 10 GW/mm^2. The time lag between the X-ray and optical pulses is controlled electronically by shifting the phase of the oscillator feedback loop with a digital delay generator. The short time jitter in the delay is on the order of a few picoseconds, but at long times it increases to tens of picoseconds. The angle between the X-ray and laser beam is 10 degrees, making the excitation geometry near collinear.

2. *Detectors*

Several detection techniques have been developed. In the first of them, the scattered X-rays are intercepted by a phosphor screen, which transforms them into optical photons. The latter are then channeled by optical fibers to a charge-coupled detector (CCD). In this detector the vast majority of X-ray photons are registered. However, it is not a straightforward method to measure very small relative changes in the CCD signal. The detection should be strictly linear in the photon number, which is not easy to achieve in practice. Finally, the X-ray dose must be kept constant during data collection. The exposure must thus be prolonged to compensate for the decaying synchrotron current. Another very different detection technique consists of using streak cameras. The incident X-ray pulse is transformed into an electronic pulse, which is "streaked" by an electrostatic field onto a CCD. Although there are streak cameras with picosecond time resolution, they require very high X-ray intensities and are therefore of limited use. There is no perfect detector, all have strengths and weaknesses, and the optimum choice depends on the exact nature of the measurements.

3. *Data Reduction Procedures*

As emphasized earlier, the weakness of the difference intensity is a specific difficulty with this sort of experiment. The ratio between the difference and the full scattering intensity is on the order of 10^{-2}–10^{-4}. This is particularly problematic in solution work, where the radiation scattered by the solute is buried in that from the solvent. A further complication arises from the interference between X-ray and optical manipulations. In fact, the intensity of

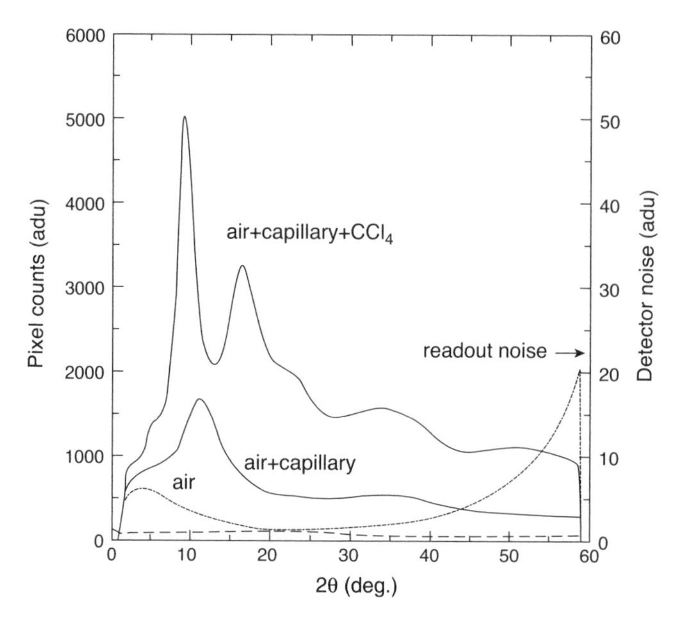

Figure 3. Various contributions to the scattered X-ray intensity. The system under consideration is a dilute I_2/CCl_4 solution.

the difference radiation depends on the number of excited molecules, which in turn is a function of the solute concentration. If the system is excited with ultrashort optical pulses, this number should not exceed a critical system-specific level. If this restriction is not respected, multiphoton absorption may be activated in the solute and solvent, and this may obscure the interpretation. In practice, the concentration of excited species is thus always very low, typically $\frac{1}{1000}$. Once again, the intensity of the X-ray beam is all important.

Another specific difficulty in X-ray experiments is that diffraction images contain contributions not only from the liquid sample but also from the capillary and air. It is often delicate to disentangle these contributions from each other (Fig. 3). Note that the main noise in the difference signal comes from the photon statistics in the X-ray background. This is an intrinsic limitation in solution phase ultrafast X-ray scattering. One must also take into account the presence of radioactivity and cosmic rays. The parasitic counts generated by these two mechanisms may be eliminated by subtracting two images, the original minus 180° rotated.

All measured intensities can be put on absolute scale by proceeding as follows. At high angles the scattering pattern can be considered as arising from a collection of noninteracting gas molecules rather than from a liquid sample. The Compton scattering cannot be neglected, but it is independent of molecular structure. Then, fitting experimental data to formulas from gas phase theory, the concentration of excited molecules can be determined. Another problem is that the undulator X-ray spectrum is not strictly monochromatic but has a slightly asymmetric lineshape extending toward lower energies. This problem may be handled in different ways, for example, by approximating its spectral distribution by its first spectral moment [12].

III. THEORY

A. Generalities

The first theoretical attempts in the field of time-resolved X-ray diffraction were entirely empirical. More precise theoretical work appeared only in the late 1990s and is due to Wilson and co-workers [13–16]. However, this theoretical work still remained preliminary. A really satisfactory approach must be statistical. In fact, macroscopic transport coefficients like the diffusion constant or the chemical rate constant break down at ultrashort time scales. Even the notion of a molecule becomes ambiguous: At which interatomic distance can the atoms A and B of a molecule A–B be considered to be free? Another element of consideration is that the electric field of the laser pump is strong and its interaction with matter is nonlinear. What is needed is thus a statistical theory reminiscent of those from time-resolved optical spectroscopy. A theory of this sort has been elaborated by Bratos and co-workers [17–19].

An important specific feature of the present experiment is worth noting. The X-ray photons have energies that are several orders of magnitude larger than those of optical photons. The pump and probe processes thus evolve on different time scales and can be treated separately. It is convenient to start with the X-ray probing processes and treat them by Maxwellian electrodynamics. The pumping processes are studied next using statistical mechanisms of nonlinear optical processes. The electron number density $n(\mathbf{r},t)$, supposed to be known in the first step, is actually calculated in this second step.

We shall now focus our attention on spatially isotropic liquids. The key quantities of the theory are as follows. An intense optical pulse of frequency Ω_O brings it into an appropriate initial state. τ seconds later, an X-ray pulse of frequency $\Omega_X \gg \Omega_O$ hits the sample and is then diffracted by it. What one measures is the difference signal $\Delta S(\mathbf{q}, \tau)$, defined as the time-integrated X-ray energy flux $S(\mathbf{q}, \tau)$ scattered in a given solid angle in the presence of the pump, minus the time-integrated X-ray energy flux $S(\mathbf{q})$ in the same angle in the absence of the pump. It depends on two variables: the scattering wavevector $\mathbf{q} = \mathbf{q}_r - \mathbf{q}_s$, where $\mathbf{q}_r$ and $\mathbf{q}_s$ are wavevectors of the incident and the scattered X-ray radiation, respectively, and the time delay τ between pump and probe. $\Delta S(\mathbf{q}, \tau)$ is the main quantity to be examined in what follows.

B. Maxwellian Description of X-ray Probing

The Maxwell theory of X-ray scattering by stable systems, both solids and liquids, is described in many textbooks. A simple and compact presentation is given in Chapter 15 of the Landau–Lifshitz volume, *Electrodynamics of Continuous Media* [20]. The incident electric and magnetic X-ray fields are plane waves $\mathbf{E}_X(\mathbf{r},t) = \mathbf{E}_{X0}\exp[i(\mathbf{q}_I\mathbf{r} - \Omega_X t)]$ and $\mathbf{H}(\mathbf{r},t) = \mathbf{H}_{X0}\exp[i(\mathbf{q}_I\mathbf{r} - \Omega_X t)]$ with a spatially and temporally constant amplitude. The electric field $\mathbf{E}_X(\mathbf{r},t)$ induces a forced oscillation of the electrons in the body. They then act as elementary antennas emitting the scattered X-ray radiation. For many purposes, the electrons may be considered to be free. One then finds that the intensity $I_X(\mathbf{q})$ of the X-ray radiation scattered along the wavevector $\mathbf{q}$ is

$$I_X(\mathbf{q}) = \left(\frac{e^2}{mc^2}\right)^2 \sin^2\phi I_{0X} f^*(\mathbf{q}) f(\mathbf{q}) \tag{1}$$

where I_{0X} is the intensity of the incident X-ray radiation, ϕ is the angle between $\mathbf{E}_X$ and $\mathbf{q}$, and $f(\mathbf{q}) = \int d\mathbf{r}\exp(-i\mathbf{q}\mathbf{r})n(\mathbf{r})$ is the Fourier-transformed electron number density $n(\mathbf{r})$; this latter quantity is generally termed a form factor. The success of this theory is immense; see the textbooks by Guinier [21], Warren [22], and Als-Nielsen and Morrow [23]. In ordered systems like crystals, it permits one to determine atomic positions, that is, to "photograph" them. In disordered systems like powders or liquids, the data are less complete but still remain very usable. A large variety of systems have been analyzed successfully using this approach.

How must this theory be modified to describe the effect of the optical excitation? The incident electric and magnetic X-ray fields are now pulses $\mathbf{E}_X(\mathbf{r},t) = \mathbf{E}_{X0}(t)\exp[i(\mathbf{q}_I\mathbf{r} - \Omega_X t)]$ and $\mathbf{H}_X(\mathbf{r},t) = \mathbf{H}_{X0}(t)\exp[i(\mathbf{q}_I\mathbf{r} - \Omega_X t)]$. They still are plane waves with a carrier frequency Ω_X, but their amplitudes $\mathbf{E}_{X0}(t)$ and $\mathbf{H}_{X0}(t)$ vary with time. The same statement applies to the electron density $n(\mathbf{r},t)$, which also is time dependent. However, these variations are all slow with time scales on the order of $1/\Omega_X$, and one can neglect $\partial\mathbf{E}_{X0}(t)/\partial t$ and $\partial\mathbf{H}_{X0}(t)/\partial t$ as compared to $i\Omega_X\mathbf{E}_{X0}(t)$ and $i\Omega_X\mathbf{H}_{X0}(t)$. Detailed calculations then show that [17]

$$S(\mathbf{q},\tau) = \left(\frac{e^2}{mc^2}\right)^2 \sin^2\phi \int_{-\infty}^{\infty} dt\, I_{0X}(t)\langle(\mathbf{q},t+\tau)f(\mathbf{q},t+\tau)\rangle \tag{2}$$

where $S(\mathbf{q},\tau)$ is time-integrated intensity of the X-ray pulse, scattered τ seconds after optical excitation in the direction of the vector $\mathbf{q}_\mathbf{r} - \mathbf{q}$. This expression can be deduced from that for $I_X(\mathbf{q})$ using the following arguments: (1) As the incident X-ray radiation is pulsed, one must replace I_{0X} by $I_{0X}(t)$. (2) Optical excitation brings the system out of thermal equilibrium. It no longer remains stationary but varies with time; thus $n(\mathbf{r}) \to n(\mathbf{r},t)$ and $f(\mathbf{q}) \to f(\mathbf{q},t) = \int d\mathbf{r}\exp(-i\mathbf{q}\mathbf{r})n$

$(\mathbf{r}, t)$. (3) The quantity $S(\mathbf{q}, \tau)$ is a time-integrated quantity. It is then useful to replace the integration variable t by the variable $t + \tau$, which permits one to introduce the time delay τ between the pump and the probe explicitly. (4) An X-ray diffraction experiment does not permit one to single out a given state of the system. Only an average $\langle\,\rangle$ over all these states can be observed.

C. Statistical Description of Optical Pumping

In the above Maxwellian description of X-ray diffraction, the electron number density $n(\mathbf{r}, t)$ was considered to be a known function of $\mathbf{r}$,t. In reality, this density is modulated by the laser excitation and is not known *a priori*. However, it can be determined using methods of statistical mechanics of nonlinear optical processes, similar to those used in time-resolved optical spectroscopy [4]. The laser-generated electric field can be expressed as $\mathbf{E}(\mathbf{r}, t) = \mathbf{E}_{\mathrm{O}0}(t)\exp(i(\mathbf{q}_{\mathrm{O}}r - \Omega_{\mathrm{O}}t))$, where Ω_{O} is the optical frequency and $\mathbf{q}_{\mathrm{O}}$ the corresponding wavevector. The calculation can be sketched as follows.

The main problem is to calculate $\langle f^*(\mathbf{q}, t+\tau)f(\mathbf{q}, t+\tau)\rangle$ of Eq. (2). To achieve this goal, one first considers $\mathbf{E}(\mathbf{r}, t)$ as a well defined, deterministic quantity. Its effect on the system may then be determined by treating the von Neumann equation for the density matrix $\rho(t)$ by perturbation theory; the laser perturbation is supposed to be sufficiently small to permit a perturbation expansion. Once $\rho(t)$ has been calculated, the quantity

$$\langle f^*(\mathbf{q}, t+\tau)f(\mathbf{q}, t+\tau)\rangle = \mathrm{Tr}[\rho(t+\tau)f^*(\mathbf{q})f(\mathbf{q})] \tag{3}$$

can be determined for a given realization of the electric field $\mathbf{E}(\mathbf{r}, t)$. In the second step, this restriction to deterministic processes is suppressed and the incident laser field $\mathbf{E}(\mathbf{r}, t)$ is identified as a stochastic quantity. In reality, $\mathbf{E}(\mathbf{r}, t)$ is never completely coherent: averaging over this stochastic process is thus necessary. This can be done using theories of transmission of electric signals. The resulting expression is inserted into Eq. (2).

D. Difference Signal $\Delta S(\mathbf{q}, \tau)$

The theoretical difference signal $\Delta S(\mathbf{q}, \tau)$ is a convolution between the temporal profile of the X-ray pulse $I_{\mathrm{OX}}(t)$ and the diffraction signal $\Delta S_{\mathrm{inst}}(\mathbf{q}, t)$ from an infinitely short X-ray pulse. This expression is [17]

$$\begin{aligned}
\Delta S(\mathbf{q}, \tau) &= \int_{-\infty}^{\infty} dt\, I_{\mathrm{OX}}(t-\tau)\Delta S_{\mathrm{inst}}(\mathbf{q}, t) \\
\Delta S_{\mathrm{inst}}(\mathbf{q}, t) &= \left(\frac{e^2}{mc^2h}\right)^2 \sin^2\theta \int_{-\infty}^{\infty}\int_{-\infty}^{\infty} d\tau_1\, d\tau_2 \\
&\quad \times \langle E_i(\mathbf{r}, t-\tau_1)E_j(\mathbf{r}, t-\tau_1-\tau_2)\rangle_{\mathrm{O}} \\
&\quad \times \langle[[f(\mathbf{q}, \tau_1+\tau_2)f^*(\mathbf{q}, \tau_1+\tau_2), M_i(\tau_2], M_j(0)]\rangle_{\mathrm{S}}
\end{aligned} \tag{4}$$

where $\mathbf{E} = (E_x, E_y, E_z)$ is the electric field of the optical pulse and $\mathbf{M} = (M_x, M_y, M_z)$ is the dipole moment of the system. Moreover, the indices i, j designate the Cartesian components x, y, z of these vectors, $\langle\rangle_{\mathrm{O}}$ realizes an averaging over all possible realizations of the optical field $\mathbf{E}$, and $\langle\rangle_{\mathrm{S}}$ realizes an averaging over the states of the nonperturbed liquid sample. Two three-time correlation functions are present in Eq. (4): the correlation function of $\mathbf{E}(t)$ and the correlation function of the variables $f(\mathbf{q}, t), \mathbf{M}(t)$. Such objects are typical for statistical mechanisms of systems out of equilibrium, and they are well known in time-resolved optical spectroscopy [4]. The above expression for $\Delta S(\mathbf{q}, \tau)$ is an exact second-order perturbation theory result.

Its general form can easily be understood. The static intensity $I_{\mathrm{X}}(\mathbf{q})$ contains the factor $f^*(\mathbf{q})f(\mathbf{q})$; its time-dependent analogue $\Delta S(\mathbf{q}, \tau)$ should then contain the factor $\langle f^*(\mathbf{q}, \tau_1 + \tau_2)f(\mathbf{q}; \tau_1 + \tau_2)\rangle$. The remaining quantities present in Eq. (4) describe optical excitation. According to Fermi's golden rule, the rate of the latter is on the order of $\sim 1/\hbar^2(\mathbf{EM}_{IF})^2$. The presence of the quantities $1/\hbar^2$, $E_i(t - \tau_1)$, $E_j(t - \tau_1 - \tau_2)$, $M_j(0)$, and $M_i(\tau_2)$ is thus natural. It should finally be noted that the scattering process depends on the properties of the material system (through $f(\mathbf{q}, t)$, $\mathbf{M}(t)$), as well as on those of the laser fields (through $\mathbf{E}(r, t)$). They determine jointly the form of the signal.

IV. LONG AND SHORT TIME LIMITS

A. Long Time Signals

The theory of time-resolved X-ray scattering has a comparatively simple limit if the optical excitation is fast compared with the process to be investigated. This so-called quasistatic condition is of great practical importance. In fact, optical pumping is generally done on subpicosecond time scales, whereas with present state-of-the-art, the X-ray probing is at least 100 times longer. A new time-scale separation appears. The slow variable is the chemically driven electron density f, and the fast variable is the laser-controlled dipole moment $\mathbf{M}$. The correlation function $\langle[[f^*f, \mathbf{M}], \mathbf{M}]\rangle$ in Eq. (4) thus splits into two factors, a factor involving f, f^* and a factor involving **M**,**M**. A quasistatic experiment thus has the intrinsic power to disentangle these two sorts of dynamics. This is the first simplification in this problem.

A second simplification results from introducing the Born–Oppenheimer separation of electronic and nuclear motions; for convenience, the latter is most often considered to be classical. Each excited electronic state of the molecule can then be considered as a distinct molecular species, and the laser-excited system can be viewed as a mixture of them. The local structure of such a system is generally described in terms of atom–atom distribution functions $g_{\mu\nu}(r, t)$ [22, 24, 25]. These functions are proportional to the probability of finding the nuclei μ and ν at the distance r at time t. Building this information into Eq. (4) and considering the isotropy of a liquid system simplifies the theory considerably.

B. "Filming" Atomic Motions

The most spectacular success of the theory in its quasistatic limit is to show how to "film" atomic motions during a physicochemical process. As is widely known, "photographing" atomic positions in a liquid can be achieved in static problems by Fourier sine transforming the X-ray diffraction pattern [22]. The situation is particularly simple in atomic liquids, where the well-known Zernicke–Prins formula provides $g(r)$ directly. Can this procedure be transfered to the quasistatic case? The answer is yes, although some precautions are necessary. The theoretical recipe is as follows: (1) Build the quantity $F(q)q\,\Delta S(q,\tau)$, where $F(q) = [\Sigma\Sigma_{\mu\neq\nu} f_\mu(q) f_\nu(q)]^{-1}$ is the "sharpening factor" and $f_\mu(q)$ and $f_\nu(q)$ are atomic form factors. (2) Perform the Fourier sine transform of this quantity for a large set of interatomic distances r. Denote the resulting signal by $\Delta S[r,\tau]$. Then, in the case of an atomic liquid [18, 19],

$$\Delta S[r,\tau] = \int_{-\infty}^{\infty} dt\, I_x(t-\tau)\Delta S_{\text{inst}}[r,t]$$

$$\Delta S_{\text{inst}}[r,t] = \left(\frac{e^2}{mc^2}\right)^2 \sin^2\theta\left[\left(\frac{1}{V(t)}g(r,t) - \frac{1}{V(0)}g(r)\right) - \left(\frac{1}{V(t)} - \frac{1}{V(0)}\right)\right] \tag{5}$$

where $g(r,t)$ is the atom pair distribution function in the presence of laser excitation, and $g(r)$ is its analogue in the laser-free system. The interpretation of the above result is as follows. (1) The first term appearing in the expression for $\Delta S[r,\tau]$ describes the variation in the pair distribution function due to laser excitation. It permits the visualization of molecular dynamics in the laser-excited system. (2) The second term probes the change in the volume $V(t)$ due to the laser heating. It dominates $\Delta S_{\text{inst}}[r,t]$ at small r's, where $g(r,t) \to 0$ and only $1/V(t) - 1/V(0)$ survives. As $\Delta V/V = -\Delta_{\rho_M/\rho_M}$, the evolution of the mass density ρ_M of the liquid can be monitored in this way. (3) Equation (5) represents a generalization of the Zernicke–Prins formula for time-resolved X-ray experiments. At small r's it permits one to monitor macroscopic variations of the mass density ρ_M, whereas at large r's it offers a visualization of atomic motions in the system.

In molecular liquids the situation is slightly more complicated; the following points merit discussion. (1) The signal $\Delta S[r,\tau]$ is composed of a number of different distribution functions $g_{\mu\nu}(r,t)$, but this is not a real handicap. If the bond μ–ν is broken, only the distribution function $g_{\mu\nu}$ of atoms μ, ν forming this bond stands out in the difference signal. Other terms disappear, partially or completely. (2) The atom–atom distribution functions $g_{\mu\nu}(r, t)$ do not enter into $\Delta S_{\text{inst}}[r, t]$ alone but are multiplied by the atomic form factors $f_\mu(q)f_\nu(q)$. This sort of coupling blurs the information contained in the signal $\Delta S[r, t]$ to a certain extent. Nevertheless, taking the necessary precautions, monitoring atomic motions is still possible.

C. Short Time Signals

Contrary to the long time limit of Eq. (4) for the signal $\Delta S(\mathbf{q}, \tau)$, its short time limit has not yet been explored in detail. The reason is that these times are not yet accessible to the experiment, due to technical difficulties. Nevertheless, some characteristics of short time signals can be understood without any detailed study. (1) First, the liquid is not isotropic at times $t \ll \tau_R$, where τ_R is molecular rotational relaxation time [13–15]. The reason is that the laser-generated electric field $\mathbf{E}(\mathbf{r}, t)$ induces a partial alignment of molecular transition moments, and of the molecules themselves in its direction. The liquid is closer to being an incompletely ordered crystal than an isotropic liquid. The difference signal $\Delta S(\mathbf{q}, \tau)$ is then expected to depend on the relative orientation of the vectors $\mathbf{q}$ and $\mathbf{E}(\mathbf{r}, t)$. The isotropy is recovered again at times $t \gg \tau_R$. (2) The quantity $\langle \exp(-i\mathbf{qr}) \rangle$, omnipresent in the theory of X-ray scattering, does not reduce to the function $\sin(qr)/qr$ in the absence of isotropy. The Zernicke–Prins formula and its extension given by Eq. (5) rely heavily on the isotropy of the liquid medium; they thus break down in the short time limit. (3) If two electronic states of the molecule are close enough to be excited simultaneously by the optical pulse, beating phenomena may occur. However, it is not known how these processes evolve on the very shortest time scales in the presence of molecular rotations. These few statements are by no means exhaustive and a detailed study of the short time regime remains to be done. However, one conclusion is certain: time-resolved X-ray diffraction is very different at long and short times.

V. SIMULATIONS AND CALCULATIONS

A. Generalities

The purpose of this section is to show how the above theory can be applied in practical calculations. For the time being, only quasistatic processes have been studied in detail; the subsequent discussion will thus focus on this limit. Two sorts of quantities enter into the theory: the atom–atom distribution functions $g^j_{\mu\nu}(r, t)$ in a given electronic state j and the corresponding populations $n_j(t)$. The total atom–atom distribution function $g_{\mu\nu}(r, t)$ is then

$$g_{\mu\nu}(r, t) = \sum_j n_j(t) g^j_{\mu\nu}(r, t) \tag{6}$$

One would prefer to be able to calculate all of them by molecular dynamics simulations exclusively. This is unfortunately not possible at present. In fact, some indices μ,ν of Eq. (6) refer to electronically excited molecules, which decay through population relaxation on the pico- and nanosecond time scales. The other indices μ,ν denote molecules that remain in their electronic ground state, and hydrodynamic time scales beyond microseconds intervene. The presence of these long times precludes the exclusive use of molecular dynamics, and a recourse to hydrodynamics of continuous media is inevitable. This concession has a high price. Macroscopic hydrodynamics assume a local thermodynamic equilibrium, which does not exist at times prior to 100 ps. These times are thus excluded from these studies.

B. Molecular Dynamics Simulations

The basic principles are described in many textbooks [24, 26]. They are thus only sketchily presented here. In a conventional classical molecular dynamics calculation, a system of N particles is placed within a cell of fixed volume, most frequently cubic in size. A set of velocities is also assigned, usually drawn from a Maxwell–Boltzmann distribution appropriate to the temperature of interest and selected in a way so as to make the net linear momentum zero. The subsequent trajectories of the particles are then calculated using the Newtonian equations of motion. Employing the finite difference method, this set of differential equations is transformed into a set of algebraic equations, which are solved by computer. The particles are assumed to interact through some prescribed force law. The dispersion, dipole–dipole, and polarization forces are typically included; whenever possible, they are taken from the literature.

Molecular dynamics permits one to determine rapidly varying atom-pair distribution functions $g^j_{\mu\nu}(r,t)$. In order to do that, one counts the number of atoms of the species ν in a spherical shell of radius r and thickness Δr centered on an atom of the species μ. This counting is repeated for a large number of computer-generated configurations. Calculations of this kind are generally well controlled, although they may occasionally generate spurious peaks in the difference signals $\Delta S(\mathbf{q})$. This perturbing effect is due to the finite size of the basic computation cell and to the smallness of difference signals. Particular care is thus necessary in the interpretation of these calculations.

The determination of the laser-generated populations $n_j(t)$ is infinitely more delicate. Computer simulations can certainly be applied to study population relaxation times of different electronic states. However, such simulations are no longer completely classical. Semiclassical simulations have been invented for that purpose, and methods such as surface hopping were proposed. Unfortunately, they have not yet been employed in the present context. Laser spectroscopic data are used instead: The decay of the excited state populations is written $n_j(t) = n_j \exp(-t/\tau_j)$, where τ_j is the experimentally determined population relaxation time. The laws of chemical kinetics may also be used when necessary. Proceeding in this way, the rapidly varying component of $\Delta S(q, \tau)$ can be determined.

C. Hydrodynamics

The calculation of the slow components in $\Delta S(q, \tau)$ follows another path. It can be realized as follows: (1) Assuming local equilibrium one can write

$$\Delta S(q, \tau) = (\delta \Delta S(q)/\delta T)_\rho \Delta T(\tau) + (\delta \Delta S(q)/\delta \rho)_T \Delta \rho(\tau) \tag{7}$$

where $\Delta T(\tau)$ and $\Delta \rho(\tau)$ are changes in the temperature and density from their equilibrium values T and ρ [18, 19]. (2) In order to estimate the thermodynamic derivatives contained in $\Delta S(q)$, this signal is calculated by molecular dynamics simulation for two temperatures T_1 and T_2 and for two densities ρ_1 and ρ_2. Finite differences are used next to calculate the derivatives. (3) The temperature and density increments $\Delta T(\tau)$ and $\Delta \rho(\tau)$ are calculated using equations of hydrodynamics of nonviscous fluids [27]. Under the present conditions, these equations can be linearized. Then, denoting by $T'(t)$ and $\rho'(t)$ the density increments $\Delta T(t)$ and $\Delta \rho(t)$, the following equations of motion can be used to calculate the time-dependent parts of $T'(t)$ and $\rho'(t)$:

$$\begin{aligned} \nabla^2 \rho' - \frac{1}{c^2} \frac{\partial^2 \rho'}{\partial t^2} &= \frac{\alpha_p}{C_p} \nabla^2 Q \\ T' &= \frac{Q}{C_p \rho_0} + \frac{C_p - C_v}{C_v} \frac{1}{\alpha_p \rho_0} \rho' \end{aligned} \tag{8}$$

Here $Q(t)$ denotes the heat input per unit volume accumulated up to time t, C_p is the specific heat per unit mass at constant pressure, C_v is the specific heat per unit mass at constant volume, c is the sound velocity, α_p is the coefficient of isobaric thermal expansion, and ρ_0 is the equilibrium density. (4) The heat input $Q(t)$ is the laser energy released by the absorbing molecule per unit volume. If the excitation is in the visible spectral range, the evolution of $Q(t)$ follows the rhythm of the different chemically driven relaxation processes through which energy is transmitted to the liquid medium. If, on the other hand, vibrations are excited in the near-infrared or infrared, the evolution of $Q(t)$ can be considered impulsive. (5) While the expression for $T'(\tau)$ is obtained immediately from $Q(\tau)$ and $\rho'(\tau)$, the calculation of $\rho'(\tau)$ is not trivial. An approximate solution, valid along the axis of a Gaussian laser beam, was found by Longacker and Litvak [28].

A slightly extended version of it is [19]

$$\rho'(\tau) = \frac{\alpha_p}{C_p} \int_{-\infty}^{\tau} dt \left(\frac{dQ(t)}{dt} \right) I - 1 + \exp(-c^2 (\tau - t)^2 / R^2))] \tag{9}$$

where R is the radius of the laser beam. The important feature of the Longaker–Litvak [28] solution is that thermal expansion does not set in immediately after a transient heat input, but is delayed as perturbations in a liquid cannot propagate faster than sound waves. There exists an acoustic horizon; the quantity $\tau_a = R/c$ is often called the acoustic transit time. This closes the discussion of the slowly varying component of the difference signal $\Delta S(q, \tau)$.

VI. RECENT ACHIEVEMENTS

A. Diffractions and Absorption

During the last few years, time-resolved X-ray techniques have been employed in several areas. The first problem studied by time-resolved X-ray diffraction was surface melting of crystals. Thermal effects consecutive to the impact of an intense laser pulse on the crystal surface were measured on picosecond and subpicosecond time scales. A number of different studies were made: melting of the crystal surface, lattice strain propagation, and onset of lattice expansion. These effects were observed for crystals such as Au [29], Ge [30], GaAs [31], and InSb [32–34]. More complex objects like Languir–Blodgett films were examined too [35]. Another very interesting research area concerns biological systems. Crystals of the myoglobin complex with CO and MbCO were examined with particular care and the positions of CO were determined at different times. Combining Laue X-ray diffraction with time-resolved optical spectroscopy permitted deeper insight into how to initiate a biochemical reaction in a crystal in a nondestructive way [36, 37]. The studies of the yellow protein in its pR state belong to the same general area [38].

Time-resolved X-ray absorption is a very different class of experiments [5–7]. Chemical reactions are triggered by an ultrafast laser pulse, but the laser-induced change in geometry is observed by absorption rather than diffraction. This technique permits one to monitor local rather than global changes in the system. What one measures in practice is the extended X-ray absorption fine structure (EXAFS), and the X-ray extended near-edge structure (XANES). Systems like SF_6 [39,40], H_2O [41], CH_3OH [41], and CB_4/C_6 [42] have been examined using this technique. Three recent papers on ruthenium(II) tris-2,2′-bipyridine, or $[Ru^{2+}(bpu)_3]^{2+}$ [43], on photosynthetic O_2 formation in biological systems [44], and on photoexcitation of NITPP – I_2 [45] in solution also merit attention. Theoretical work advanced at the same time. Early approaches are due to Frank et al. [46], whereas a statistical theory of time-resolved X-ray absorption was proposed by Tanaka et al. [47] and Tanaka and Mukamel [48]. This latter theory represents the counterpart of the X-ray diffraction theory developed in this chapter.

The purpose of this section is to describe recent achievements in time-resolved X-ray diffraction from liquids. Keeping the scope of the present chapter in mind, neither X-ray diffraction from solids nor X-ray absorption will be discussed. The majority of experiments realized up till now were performed using optical excitation, although some recent attempts using infrared excitation were also reported. The main topics that have been studied are (1) visualization of atomic motions during a chemical reaction, (2) structure of reaction intermediates in a complex reaction sequence, (3) heat propagation in impulsively heated liquids, and (4) chemical hydrodynamics of nanoparticle suspensions. We hope that the actual state-of-the-art will be illustrated in this way.

B. Visualization of Atomic Motions

The first reaction "filmed" by X-rays was the recombination of photodissociated iodine in a CCl_4 solution [18, 19, 49]. As this reaction is considered a prototype chemical reaction, a considerable effort was made to study it. Experimental techniques such as linear [50–52] and nonlinear [53–55] spectroscopy were used, as well as theoretical methods such as quantum chemistry [56] and molecular dynamics simulation [57]. A fair understanding of the dissociation and recombination dynamics resulted. However, a fascinating challenge remained: to "film" atomic motions during the reaction. This was done in the following way.

A dilute T_2/CC_4 solution was pumped by a 520 nm visible laser pulse, promoting the iodine molecule from its ground electronic state X to the excited states A, A', B, and ${}^1\pi_u$ (Fig. 4). The laser excited T_2 dissociates rapidly into an

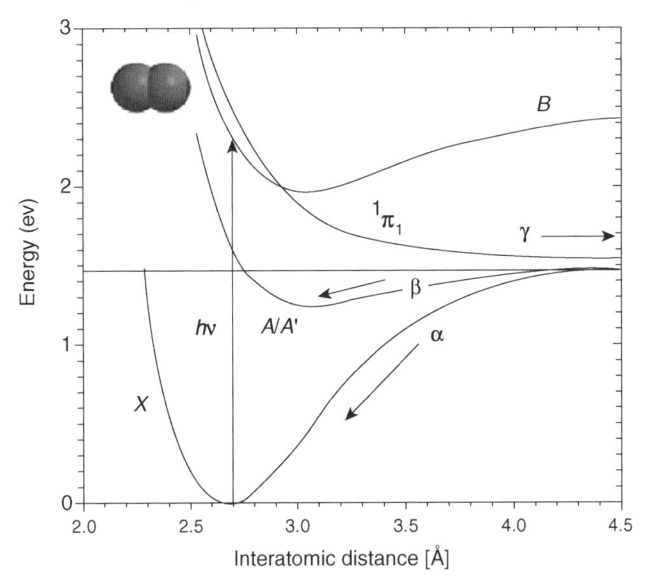

Figure 4. Low lying electronic energy surfaces of I_2. These states are termed X, A/A', B, and ${}^1\pi_u$ states. The processes α, β, and γ denote vibrational cooling along the X potential, geminate recombination through the states A/A', and nongeminate recombination, respectively.

unstable intermediate $(T_2)^*$. The latter decomposes, and the two iodine atoms recombine either geminately (a) or nongeminately (b):

$$I_2 + h\nu \rightarrow I_2^*, I_2^* \rightarrow I_2(a), I_2^* \rightarrow 2I \rightarrow I_2(b)$$

The resulting atomic motions were probed by X-ray pulses for a number of time delays. A collection of diffraction patterns were then transformed into a series of real-space snapshots; theory is required to accomplish this last step. When the sequence of snapshots were joined together, it became a film of atomic motions during recombination.

The detailed description of the "film" is as follows. The first minimum in $\Delta S[r, \tau]$ at 2.7 Å at early times is due to the depletion of the X state of molecular iodine from the laser excitation (Fig. 5). The excited molecules then reach the

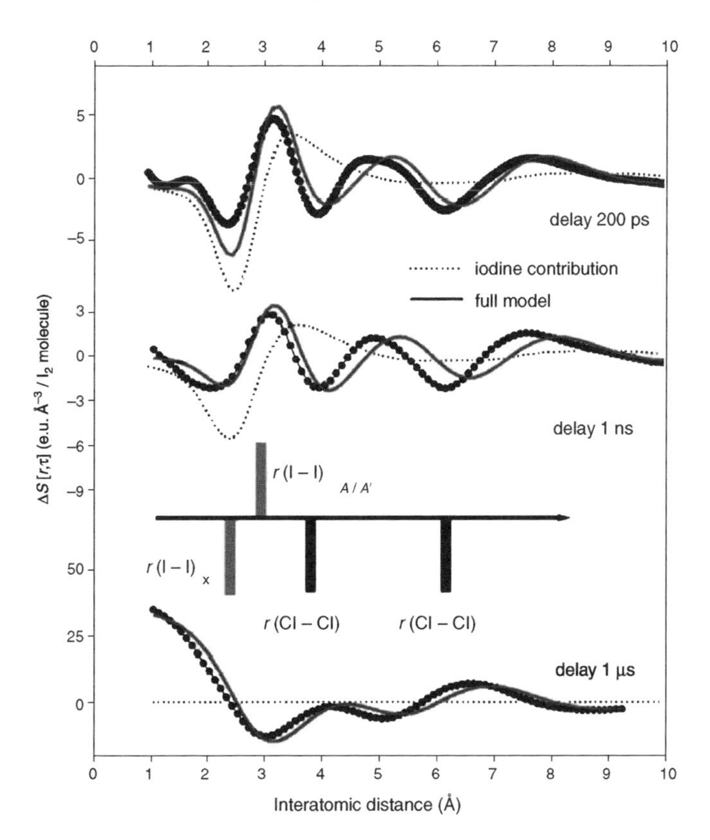

Figure 5. The Fourier transformed signal $\Delta S[r, \tau]$ of I_2/CCl_4. The pump-probe delay times are $\tau = 200$ ps, 1 ns, and 1 μs. The light gray bars indicate the bond lengths of iodine in the X and A/A' states. The dark gray bars show the positions of the first two intermolecular peaks in the pair distribution function g_{C-Cl}.

A/A' and higher electronic states, and a maximum appears around 3.2 Å. In addition, the energy transfer from the solute to the solvent induces a rearrangement of the structure of the liquid without any observable thermal expansion. New minima in $\Delta S[r, \tau]$ appear at 4.0 and 6.2 Å. Molecular dynamics simulations assigned them to changes in the intermolecular Cl–Cl distances in liquid CCl_4. At later times, the excited molecules all relax, and the thermal expansion is completed. The strong increase in $\Delta S[r, \tau]$ observed at small r's is due to the decrease in the mass density ρ_M of CCl_4, compared with Eq. (5). In turn, the features observed at large r's reflect the variations in the intermolecular Cl–Cl distances due to the thermal expansion.

"Filming" of atomic motions in liquids was thus accomplished. More specifically, the above experiment provides atom–atom distribution functions $g_{\mu\nu}(r, \tau)$ as they change during a chemical reaction. It also permits one to monitor temporal variations in the mean density of laser-heated solutions. Last but not least, it shows that motions of reactive and solvent molecules are strongly correlated: the solvent is not an inert medium hosting the reaction [58].

C. Structure of Reaction Intermediates

1. *Photodissociation of Diiodomethane*

Determining the geometry of short lived transients in a complex chemical reaction is a difficult—but very important—problem of chemistry. Time-resolved X-ray diffraction offers new possibilities in this domain, as shown by a recent work on photodissociation of diiodomethane CH_2I_2 in methanol CH_3OH [59]. For many reasons, partially scientific and partially commercial, this reaction was extensively studied in the past. Spectroscopic techniques [60–63] were employed, together with theoretical methods of quantum chemistry [64–66]. In spite of these efforts, the reaction mechanism remained poorly understood. For example, Tarnovsky [63] suggested the presence of a long living intermediate (CH_2I—I) in the reaction sequence. In order to prove—or disprove—this statement, the problem was reexamined by time-resolved X-ray diffraction.

The experiment was as follows. A diluted solution of CH_2I_2 was pumped by an optical laser, promoting an electron onto an antibonding orbital of the C—I bond. After excitation this bond is ruptured and the $(I)^*$ and $(CH_2I)^*$ radicals are formed. Several reaction pathways are possible, both geminate and nongeminate. These radicals also react with CH_3OH to form the ions I_2^- and I_3^-. The solution was proved by time-delayed 150 ps long X-ray pulses.

The theory was very similar to that described earlier but was simplified in view of the complexity of the problem. A number of reaction intermediates were considered explicitly, and the corresponding signals were calculated by molecular dynamics simulation. Kinetic equations governing the reaction sequence were established and were solved numerically. The main simplification of the theory is that, when calculating $\Delta S[r,\tau]$, the lower limit of the Fourier integral was shifted from 0 to a small value q_M. The authors wrote [59]

$$\Delta S[r,\tau] = \int_0^\infty dq\, qF(q)\Delta S(q,\tau)\sin qr \rightarrow \int_{q_M}^\infty dq\, qF(q)\Delta S(q,\tau)\sin qr \quad (10)$$

The rationale of this assumption is that low-q contributions to $\Delta S[r,\tau]$ affect this quantity only at large r's. As a consequence, the low-r region, where the reaction intermediates have their signatures, should be solvent free.

The results are presented in Fig. 6, where the measured and calculated difference signals $\Delta S[r,\tau]$ are illustrated. The presence of a strong peak at

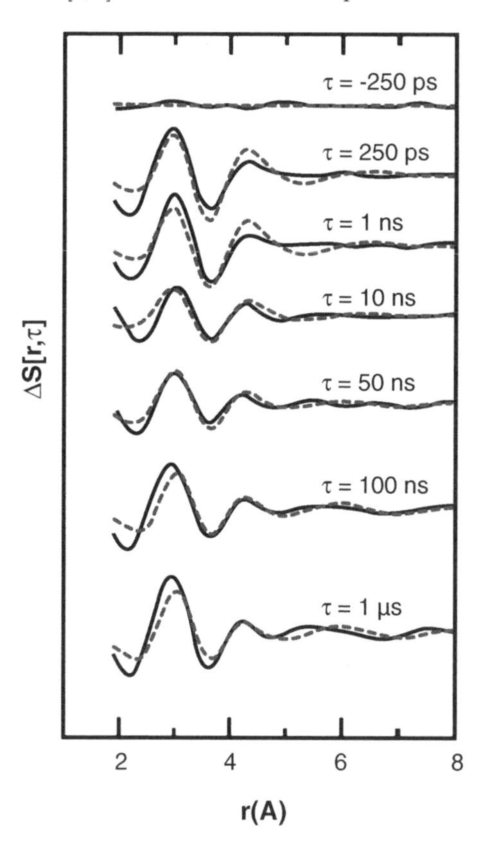

Figure 6. The Fourier transformed signal $\Delta S[r,\tau]$ of CH_2I_2/CH_3OH. The pump-probe time delays vary between $\tau = -250$ ps and 1 μs. The pair distribution function g_{I-I} peaks in the 3 Å region. If $\tau < 50$ ns, the I—I bond corresponds to the short-lived intermediate (CH_2I—I), and if $\tau > 100$ ns it belongs to the $(I_s^-)^*$ ion. Dashed curves indicate the theory, and black curves describe the experiment.

$r \approx 2.7$–3.0 Å proves the existence of an T_2 bond in different chemical environments; this bond is present in (CH_2I—I) at times less than 10 nanoseconds, and in I_3^- at later times. This study proves, for the first time unambiguously, the presence of the long living intermediate (CH_3I—I).

2. *Photodissociation of Diiodoethane*

Another chemical reaction studied with the same method was the photodissociation of diiodoethane $C_2H_4T_2$ in CH_3OH [67]. Attention was focused on the radical $(CH_2ICH_2)^*$, which appears after release of one iodine atom. This radical was examined carefully by theory [68–70] and experiment [71–74]. In spite of these efforts, the geometrical information remained incomplete. A bridged structure was postulated by Skell and co-workers to explain the stereochemistry of free radical addition reactions [75, 76], but direct structural evidence was still lacking. Should the anticonfiguration $(ICH_2\text{—}CH_2)^*$ really be excluded? The purpose of this work was to provide the missing information by time-resolved X-ray diffraction.

The experiment was done as follows. A diluted solution of $C_2H_4I_2$ in CH_3OH was pumped by an optical laser, which triggered the elimination of one iodine atom followed by creation of the radicals $(C_2H_4I)^*$ and $(I)^*$. A series of X-ray diffraction patterns were recorded at times between $-100\,\text{ps}$ and $3\,\mu\text{s}$. The signals $\Delta S(q, \tau)$ and their Fourier transforms $\Delta S[r, \tau]$ were determined in this experiment. The theory was similar to that described earlier. In particular, the solvent-free signals $\Delta S[r, \tau]$ were calculated using Eq. (10). The main difficulty in this study was that the signatures of the bridge and the antiradicals are not very different. Molecular dynamics simulations were thus realized with particular care, and statistical deviation checks (χ^2) were performed.

The results of these investigations are presented in Fig. 7, where the signals $\Delta S[r, \tau]$ are plotted for the bridge and antiform of the radical $(C_2H_2I)^*$. These figures favor the bridge form. The concentrations of different species in the solution at different times were also determined. The crucial role of theory should be emphasized: it would be difficult to extract this information by simple insight of the experimental data.

D. Hydrodynamics of Laser-Heated Liquids

If the system under consideration is chemically inert, the laser excitation only induces heat, accompanied by density and pressure waves. The excitation can be in the visible spectral region, but infrared pumping is also possible. In the latter case, the times governing the delivery of heat to the liquid are those of vibrational population relaxation. They are very short, on the order of 1 ps; this sort of excitation is thus impulsive. Contrary to first impression, the physical reality is in fact quite subtle. The acoustic horizon, described in Section V.C is in the center of the discussion [18, 19]. As laser-induced perturbations cannot propagate faster

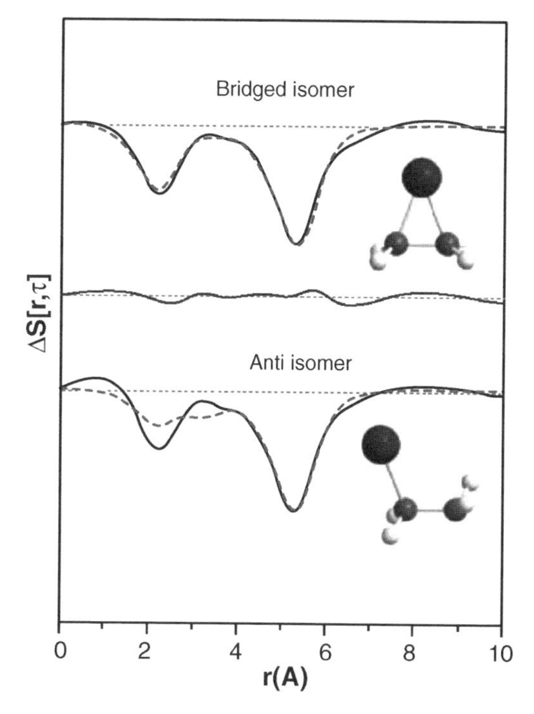

Figure 7. The Fourier transformed signal $\Delta S[r, \tau]$ of the $(C_2H_4I)^*$ radical in methanol at $\tau = 100$ ps. The agreement between theory (dashed curve) and experiment (black curve) is better if the radical is assumed to be bridged (top) rather than to have an antiform (bottom).

than sound, thermal expansion is delayed at short times. The physicochemical consequences of this delay are still entirely unknown. The liquids submitted to investigation are water and methanol.

A first study refers to liquid water [77]. The signals $\Delta S(q, \tau)$ and $\Delta S[r, \tau]$ were measured using time-resolved X-ray diffraction techniques with 100 ps resolution. Laser pulses at 266 and 400 nm were employed. Only short times τ were considered, where thermal expansion was assumed to be negligible and the density ρ to be independent of τ. To prove this assumption, the authors compared their values of $\Delta S(q, \tau)$ to the values of $\Delta S(q)$ obtained from isochoric (i.e., $\rho = \text{const}$) temperature differential data [78–80]. Their argument is based on the fact that liquid H_2O shows a density maximum at 4 °C. Pairs of temperatures T_1, T_2 thus exist for which the density ρ is the same: constant density conditions can thus be created in this unusual way. The experiment confirmed the existence of the acoustic horizon (Fig. 8).

Another study refers to liquid CH_3OH, where an infrared excitation was employed for the first time [81]. This excitation was realized using the overtone

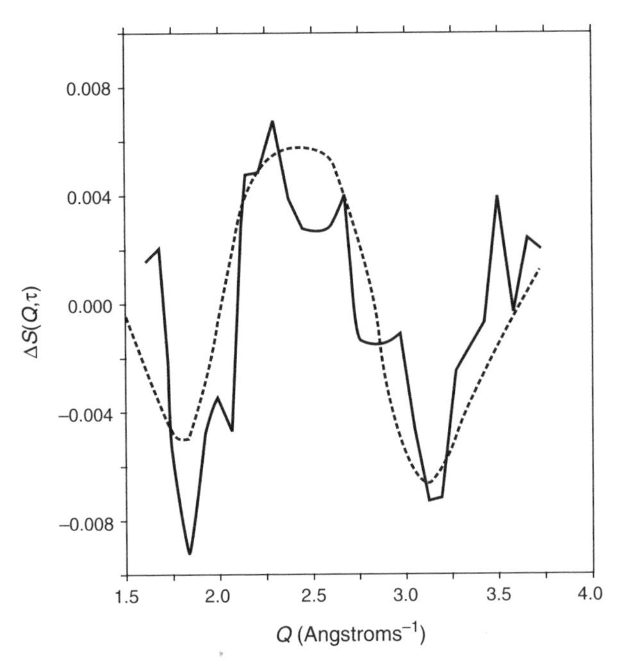

Figure 8. Comparison of difference signal $\Delta S(q, \tau)$ at $\tau = 700$ ps (solid) with static isochoric data (dashed) for water. The data were collected using a time-resolved avalanche photodiode without area sensitivity.

of the νOH mode of methanol and the asymmetric νCH mode. The time was varied over a wide range, between 100 ps and 1 μs. The time-resolved X-ray diffraction techniques employed were similar to those mentioned earlier. The authors then succeeded in extracting the difference signals $\Delta S(q, \tau)$ and in deducing the thermodynamic derivatives of the signal $(\partial \Delta S/\partial T)_\rho$ and $(\partial \Delta S/\partial \rho)_T$. These quantities, normally obtainable only by computer simulation, were employed to refine the analysis of the diiodoethane photodissociation, as discussed in Section VI.C [67]. The quality of global fits was improved considerably by employing these measured solvent differentials.

E. Gold Nanoparticles in Water

The last subject where time-resolved X-ray diffraction techniques proved their exceptional potential concerns chemical physics of gold nanoparticles in water. Belonging jointly to X-ray and nanoparticle physics, this subject is in a certain sense on the borderline of the present chapter [82–84]. Nevertheless, the possibilities offered by time-resolved X-ray techniques in this domain are fascinating. The heart of the problem is as follows. A suspension of gold nanoparticles is submitted to a laser excitation. If the excitation power is low enough, these particles are not damaged by the radiation. If not, the nanoparticles may be transformed or even destroyed. Three sorts of problems were analyzed. The first is laser-induced heating and melting after excitation with femtosecond laser pulses [85]. At lower excitation power, the lattice heating is followed by cooling on the nanosecond time scale. The lattice expansion rises linearly with the laser excitation up to a lattice temperature increase ΔT of the order of 500 K. At higher temperatures, the long range order decreases due to premelting of the particles. At still higher temperatures, complete melting occurs within the first 100 ps after laser excitation.

A second problem in these studies concerns cavitation dynamics on the nanometer length scale [86]. If sufficiently energetic, the ultrafast laser excitation of a gold nanoparticle causes strong nonequilibrium heating of the particle lattice and of the water shell close to the particle surface. Above a threshold in the laser power, which defines the onset of homogeneous nucleation, nanoscale water bubbles develop around the particles, expand, and collapse again within the first nanosecond after excitation (Fig. 9). The size of the bubbles may be examined in this way.

The last problem of this series concerns femtosecond laser ablation from gold nanoparticles [87]. In this process, solid material transforms into a volatile phase initiated by rapid deposition of energy. This ablation is nonthermal in nature. Material ejection is induced by the enhancement of the electric field

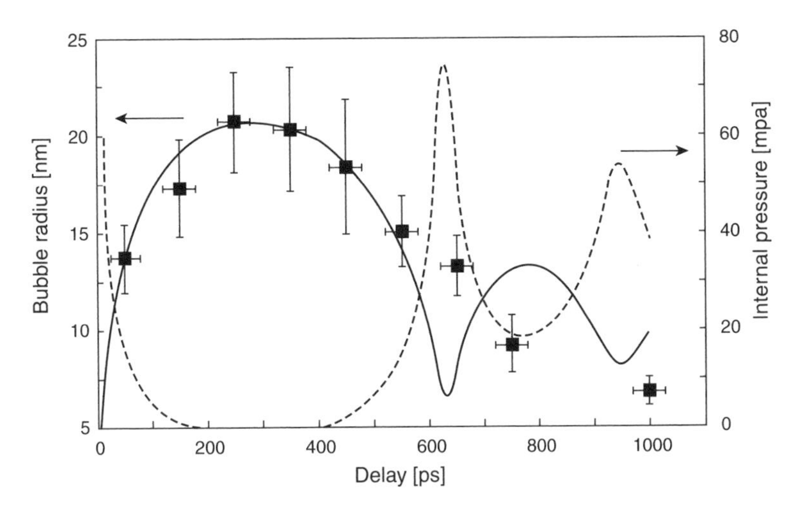

Figure 9. Bubble radius and pressure transients of the water vapor inside the bubbles. The first maximum in pressure at 650 ps marks the collapse of the bubbles. The following modulations are only expected for oscillatory bubble motion.

close to the curved nanoparticle surface. This ablation is achievable for laser excitation powers far below the onset of general catastrophic material deterioration, such as plasma formation or laser-induced explosive boiling. Anisotropy in the ablation pattern was observed. It coincides with a reduction of the surface barrier from water vaporization and particle melting. This effect limits any high power manipulation of nanostructured surfaces such as surface-enhanced Raman measurements or plasmonics with femtosecond pulses.

VII. CONCLUSIONS

A major breakthrough has been realized in the last few years in time-resolved X-ray diffraction and absorption. It may be compared to that realized fifteen years ago in time-resolved optical spectroscopy. It offers spatial and temporal resolution of atomic motions directly, contrary to optical spectroscopy where extra information is required to pass from energy to geometry. First, molecular "films" were realized and the geometry of a few short-lived reaction intermediates was determined. Surface melting of crystals, laser-induced deterioration or destruction of gold nanoparticles in water, and migration of carbon monoxide molecules in the myoglobin crystals are further examples of successful applications of this technique. A new avenue is open in this direction.

However, time-resolved X-ray diffraction remains a young science. It is still impossible, or is at least very difficult, to attain time scales of less than a picosecond. General characteristics of subpicosecond X-ray diffraction and absorption are hardly understood. To progress in this direction, free-electron laser X-ray sources are actually under construction, subject to heavy financial constraints. Nevertheless, this field is exceptionally promising. Working therein is a challenge for everybody!

Acknowledgments

The authors are indebted to Hyotcherl Ihee, Philip Anfinrud, Friederich Schotte, Anton Plech, Maciej Lorenc, Quingyu Kong, Marco Camarata, Rodolphe Vuilleumier, and Fabien Mirloup for experimental and theoretical assistance. They would also like to thank Marie-Claire Bellissent-Funel and Yan Gauduel for their help at early stages of this research. The EU grant FP6-503641 titled FLASH is also gratefully acknowledged.

References

1. Y. R. Shen, *The Principles of Nonlinear Optics*, John Wiley & Sons, Hoboken, NJ, 2002.
2. Y. Amnon, *Quantum Electronics*, 3rd ed., John Wiley & Sons, Hoboken, NJ, 1988.
3. D. M. Levenson and S. S. Kano, *Introduction to Nonlinear Laser Spectroscopy*, Optics and Photonics Series, Academic Press, New York, 1988; revised edition, 1989.
4. S. Mukamel, *Principles of Nonlinear Optical Spectroscopy*, Oxford University Press, New York, 1999.

5. J. R. Helliwell and P. M. Rentzepis, *Time-Resolved Diffraction*, Oxford Series on Synchrotron Radiation, No. 2, Oxford University Press, Oxford, UK, 1997.
6. C. Bressler and M. Chergui, Ultrafast X-ray absorption spectroscopy. *Chem. Rev.* **104**(4), 1781–1812 (2004).
7. V. I. Tomov, A. D. Oulianov, P. Chen, and M. P. Rentzepis, Ultrafast time-resolved transient structures of solids and liquids studied by means of X-ray diffraction and EXAFS. *J. Phys. Chem. B* **103**(34), 7081–7091 (1999).
8. F. Schotte, S. Techert, P. Anfinrud, V. Srajer, K. Moffat, and M. Wulff, Picosecond structural studies using pulsed synchrotron radiation, in *Third-Generation Hard X-Ray Synchrotron Radiation Sources: Source Properties, Optics, and Experimental Techniques* (D. M. Mills, ed.), John Wiley & Sons, Hoboken, NJ, 2002, Chap. 10, pp. 345–402.
9. M. Dohlus and T. Limberg, Calculation of coherent synchrotron radiation in the ttf-fel bunch compressor magnet chicanes. *Nucl. Instrum. Methods Phys. Res. Sect. A* **407**(1-3), 278–284 (1998).
10. R. W. Schoenlein, S. Chattopadhyay, H. H. W. Chong, T. E. Glover, P. A. Heimann, C. V. Shank, A. A. Zholents, and M. S. Zolotorev, Generation of femtosecond pulses of synchrotron radiation. *Science* **287**(5461), 2237–2240 (2000).
11. M. S. Gruner and H. D. Bilderback, Energy recovery linacs as synchrotron light sources. *Nucl. Instrum. Methods Phys. Res. Sec. A* **500**(1-3), 25–32 (2003).
12. A. Plech, R. R. A. Geis, and M. Wulff, Diffuse scattering from liquid solutions with white-beam undulator radiation for photoexcitation studies. *J. Synchrotron Radiat.* **9**(5), 287–292 (2002).
13. M. Ben-Nun, C. P. J. Barty, T. Guo, F. Ràksi, Ch. Rose-Petruck, J. Squier, K. R. Wilson, V. V. Yakovliev, P. M. Weber, Z. Jiang, A. Ikhlef, and J-Cl. Kieffer, Ultrafast X-ray diffraction and absorption, in *Time-Resolved Diffraction*, Vol. 2 (J. R. Helliwell and P. M. Rentzepis, eds.), Oxford Series on Synchrotron Radiation, Oxford University Press, Oxford, UK, 1997, Chap. 2, pp. 44–70.
14. M. Ben-Nun, J. Cao, and K. R. Wilson, Ultrafast X-ray and electron diffraction: theoretical considerations. *J. Phys. Chem. A* **101**(47), 8743–8761 (1997).
15. J. Cao and K. R. Wilson, Ultrafast X-ray diffraction theory. *J. Phys. Chem. A.* **102**(47), 9523–9530 (1998).
16. C. H. Chao, S. H. Lin, W. K. Liu, and R. Rentzepis, Theory of ultrafast time-resolved X-ray and electron diffraction, in *Time-Resolved Diffraction*, Vol. 2 (J. R. Helliwell and P. M. Rentzepis, eds.) Oxford University Press, Oxford Series on Synchrotron Radiation. Oxford, UK, 1997, Chap. 11, pp. 260–283.
17. S. Bratos, F. Mirloup, R. Vuilleumier, and M. Wulff, Time-resolved X-ray diffraction: statistical theory and its application to the photo-physics of molecular iodine. *J. Chem. Phys.* **116**(24), 10615–10625 (2002).
18. S. Bratos, F. Mirloup, R. Vuilleumier, M. Wulff, and A. Plech, X-ray "filming" of atomic motions in chemical reactions. *Chem. Phys.* **304**(3), 245–251 (2004).
19. M. Wulff, S. Bratos, A. Plech, R. Vuilleumier, F. Mirloup, M. Lorenc, Q. Kong, and H. Ihee, Recombination of photodissociated iodine: a time-resolved X-ray-diffraction study. *J. Chem. Phys.* **124**(3), 034501 (2006).
20. L. D. Landau, E. M. Lifshitz, and L. P. Pitaevskii, *Electrodynamics of Continuous Media*, Vol. 8, Course of Theoretical Physics, Pergamon Press, Oxford, UK, 1984.
21. A. Guinier, *X-ray Diffraction in Crystals, Imperfect Crystals, and Amorphous Bodies*, Dover, New York, 1963.

22. B. E. Warren, *X-ray Diffraction*, Dover, New York, 1990.

23. J. Als-Nielsen and D. McMorrow, *Elements of Modern X-Ray Physics*, John Wiley & Sons, Hoboken, NJ, 2000.

24. J.-P. Hansen and I. R. McDonald, *Theory of Simple Liquids*, Academic Press, London, 1986.

25. A. H. Narten and H. A. Levy, Observed diffraction pattern and proposed models of liquid water. *Science* **165**(3892), 447–454 (1969).

26. M. P. Allen and D. J. Tidesley, *Computer Simulation of Liquids*, Oxford University Press, Oxford, UK, 1989.

27. L. D. Landau, E. M. Lifshitz, and L. P. Pitaevskii, *Fluid Mechanics*, Vol. 8, Course of Theoretical Physics, Pergamon Press, Oxford, UK, 1987.

28. P. R. Longaker and M. M. Litvak, Perturbation of the refractive index of absorbing media by a pulsed laser beam. *J. Appl. Phys.* **40**(10), 4033–4041 (1969).

29. P. Chen, I. V. Tomov, and P. M. Rentzepis, Time resolved heat propagation in a gold crystal by means of picosecond X-ray diffraction. *J. Chem. Phys.* **104**(24), 10001–10007 (1996).

30. C. W. Siders, A. Cavalleri, K. Sokolowski-Tinten, Cs. Tóth, T. Guo, M. Kammler, M. Horn von Hoegen, K. R. Wilson, D. von der Linde, and C. P. J. Barty, Detection of nonthermal melting by ultrafast X-ray diffraction. *Science* **286**(5443), 1340–1342 (1999).

31. C. Rose-Petruck, R. Jimenez, T. Guo, A. Cavalleri, C. W. Siders, F. Raksi, J. A. Squier, B. C. Walker, K. R. Wilson, and C. P. J. Barty, Picosecond-milliangström lattice dynamics measured by ultrafast X-ray diffraction. *Nature* **398**, 310–312 (1999).

32. A. H. Chin, R. W. Schoenlein, T. E. Glover, P. Balling, W. P. Leemans, and C. V. Shank, Ultrafast structural dynamics in InSb probed by time-resolved X-ray diffraction. *Phys. Rev. Lett.* **83**, 336–339 (1999).

33. D. A. Reis, M. F. DeCamp, P. H. Bucksbaum, R. Clarke, E. Dufresne, M. Hertlein, R. Merlin, R. Falcone, H. Kapteyn, M. M. Murnane, J. Larsson, Th. Missalla, and J. S. Wark, Probing impulsive strain propagation with X-ray pulses. *Phys. Rev. Lett.* **86**, 3072–3075 (2001).

34. A. Rousse, C. Rischel, S. Fourmaux, I. Uschmann, S. Sebban, G. Grillon, Ph. Balcou, E. Forster, J. P. Geindre, P. Audebert, J. C. Gauthier, and D. Hulin, Non-thermal melting in semiconductors measured at femtosecond resolution. *Nature* **410**, 65–68 (2001).

35. C. Rischel, A. Rousse, I. Uschmann, P.-A. Albouy, J.-P. Geindre, P. Audebert, J.-C. Gauthier, E. Froster, J.-L. Martin, and A. Antonetti, Femtosecond time-resolved X-ray diffraction from laser-heated organic films. *Nature* **390**, 490–492 (1997).

36. V. Srajer, T. Teng, T. Ursby, C. Pradervand, Z. Ren, S. Adachi, W. Schildkamp, D. Bourgeois, M. Wulff, and K. Moffat, Photolysis of the carbon monoxide complex of myoglobin: nanosecond time-resolved crystallography. *Science* **274**, 1726–1729 (1996).

37. F. Schotte, M. Lim, T. A. Jackson, A. V. Smirnov, J. Soman, J. S. Olson, Jr., G. N. Phillips, M. Wulff, and P. A. Anfinrud, Watching a protein as it functions with 150-ps time-resolved X-ray crystallography. *Science* **300**(5627), 1944–1947 (2003).

38. B. Perman, V. Srajer, Z. Ren, Tsu yi Teng, C. Pradervand, T. Ursby, D. Bourgeois, F. Schotte, M. Wulff, R. Kort, K. Hellingwerf, and K. Moffat, Energy transduction on the nanosecond time scale: early structural events in a xanthopsin photocycle. *Science* **279**, 1946–1950 (1998).

39. F. Raksi, K. R. Wilson, Z. Jiang, A. Ikhlef, C. Y. Cote, and J.-C. Kieffer, Ultrafast X-ray absorption probing of a chemical reaction. *J. Chem. Phys.* **104**(15), 6066–6069 (1996).

40. H. Nakamatsu, T. Mukoyama, and H. Adachi, Theoretical X-ray absorption spectra of SF_6 and H_2S. *J. Chem. Phys.* **95**(5), 3167–3174 (1991).

41. K. R. Wilson, R. D. Schaller, D. T. Co, R. J. Saykally, B. S. Rude, T. Catalano, and J. D. Bozek, Surface relaxation in liquid water and methanol studied by X-ray absorption spectroscopy. *J. Chem. Phys.* **117**(16), 7738–7744 (2002).
42. I. V. Tomov and P. M. Rentzepis, Ultrafast X-ray determination of transient structures in solids and liquids. *Chem. Phys.* **299**(2-3), 203–213 (2004).
43. M. Saes, C. Bressler, R. Abela, D. Grolimund, S. L. Johnson, P. A. Heimann, and M. Chergui, Observing photochemical transients by ultrafast X-ray absorption spectroscopy. *Phys. Rev. Lett.* **90**(4), 047403 (2003).
44. M. Haumann, P. Liebisch, C. Muller, M. Barra, M. Grabolle, and H. Dau, Photo-synthetic O_2 formation tracked by time-resolved X-ray experiments. *Science* **310**(5750), 1019–1021 (2005).
45. L. X. Chen, W. J. H. Jager, G. Jennings, D. J. Gosztola, A. Munkholm, and J. P. Hessler, Capturing a photoexcited molecular structure through time-domain X-ray absorption fine structure. *Science* **292**(5515), 262–264 (2001).
46. F. L. H. Brown, K. R. Wilson, and J. Cao, Ultrafast extended X-ray absorption fine structure (EXAFS)—theoretical considerations. *J. Chem. Phys.* **111**(14), 6238–6246 (1999).
47. S. Tanaka, V. Chernyak, and S. Mukamel, Time-resolved X-ray spectroscopies: nonlinear response functions and Liouville-space pathways. *Phys. Rev. A: Atomic Mol. Opt. Phys.* **63**(6), 063405 (2001).
48. S. Tanaka and S. Mukamel, X-ray four-wave mixing in molecules. *J. Chem. Phys.* **116**(5), 1877–1891 (2002).
49. A. Plech, M. Wulff, S. Bratos, F. Mirloup, R. Vuilleumier, F. Schotte, and P. A. Anfinrud, Visualizing chemical reactions in solution by picosecond X-ray diffraction. *Phys. Rev. Lett.* **92**(12), 125505 (2004).
50. J. Franck and E. Rabinowitsch, Some remarks about free radicals and the photo-chemistry of solutions. *Trans. Faraday Soc.* **30**, 130 (1934).
51. E. Rabinowitch and W. C. Wood, Properties of illuminated iodine solutions. J. Photochemical dissociation of iodine molecules in solution. *Trans. Faraday Soc.* **32**, 547–555 (1936).
52. J. Zimmerman and R. M. Noyes, The primary quantum yield of dissociation of iodine in hexane solution. *J. Chem. Phys.* **18**(5), 658–666 (1950).
53. G. W. Hoffman, T. J. Chuang, and K. B. Eisenthal, Picosecond studies of the cage effect and collision induced predissociation of iodine in liquids. *Chem. Phys. Lett.* **25**(2), 201–205 (1974).
54. N. A. Abul-Haj and D. F. Kelley, Geminate recombination and relaxation of molecular iodine. *J. Chem. Phys.* **84**(3), 1335–1344 (1986).
55. A. L. Harris, J. K. Brown, and C. B. Harris, The nature of simple photodissociation reactions in liquids on ultrafast time scales. *Annu. Rev. Phys. Chem.* **39**, 341–366 (1988).
56. R. S. Mulliken, Iodine revisited. *J. Chem. Phys.* **55**(1), 288–309 (1971).
57. J. P. Bergsma, M. H. Coladonato, P. M. Edelsten, J. D. Kahn, K. R. Wilson, and D. R. Fredkin, Transient X-ray scattering calculated from molecular dynamics. *J. Chem. Phys.* **84**(11), 6151–6160 (1986).
58. S. A. Rice, Atom tracking. *Nature* **429**, 255–256 (2004).
59. J. Davidsson, J. Poulsen, M. Cammarata, P. Georgiou, R. Wouts, G. Katona, F. Jacobson, A. Plech, M. Wulff, G. Nyman, and R. Neutze, Structural determination of a transient isomer of CH_2I_2 by picosecond X-ray diffraction. *Phys. Rev. Lett.* **94**(24), 245503 (2005).
60. J. Zhang and D. G. Imre, CH_2I_2 photodissociation: emission spectrum at 355 nm. *J. Chem. Phys.* **89**(1), 309–313 (1988).

61. B. J. Schwartz, J. C. King, J. Z. Zhang, and C. B. Harris, Direct femtosecond measurements of single collision dominated geminate recombination times of small molecules in liquids. *Chem. Phys. Lett.* **203**(5-6), 503–508 (1993).

62. W. M. Kwok and D. L. Phillips, Solvation effects and short-time photodissociation dynamics of CH_2I_2 in solution from resonance Raman spectroscopy. *Chem. Phys. Lett.* **235** (3-4), 260–267 (1995).

63. A. N. Tarnovsky, V. Sundstrom, E. Akesson, and T. Pascher, Photochemistry of diiodomethane in solution studied by femtosecond and nanosecond laser photolysis. Formation and dark reactions of the CH_2I—I isomer photoproduct and its role in cyclopropanation of olefins. *J. Phys. Chem. A* **108**(2), 237–249 (2004).

64. M. Odelius, M. Kadi, J. Davidsson, and A. N. Tarnovsky, Photodissociation of diiodomethane in acetonitrile solution and fragment recombination into iso-diiodomethane studied with *ab initio* molecular dynamics simulations. *J. Chem. Phys.* **121**(5), 2208–2214 (2004).

65. A. E. Orel and O. Kühn, Cartesian reaction surface analysis of the CH_2I_2 ground state isomerization. *Chem. Phys. Lett.* **304**(3-4), 285–292 (1999).

66. D. L. Phillips, W.-H. Fang, and X. Zheng, Isodiiodomethane is the methylene transfer agent in cyclopropanation reactions with olefins using ultraviolet photolysis of diiodomethane in solutions: a density functional theory investigation of the reactions of isodiiodomethane, iodimethyl radical, and iodomethyl cation with ethylene, *J. Am. Chem. Soc.* **123**(18), 4197–4203 (2001).

67. H. Ihee, M. Lorenc, T. K. Kim, Q. Y. Kong, M. Cammarata, J. H. Lee, S. Bratos, and M. Wulff, Ultrafast X-ray diffraction of transient molecular structures in solution. *Science* **309**(5738), 1223–1227 (2005).

68. B. Engels and S. D. Peyerimhoff, Theoretical study of the bridging in ß-halo ethyl. *J. Mol. Structure THEOCHEM* **138**(1-2), 59–68 (1986).

69. F. Bernardi and J. Fossey, An *ab initio* study of the structural properties of ß-substituted ethyl-free radicals. *J. Mol. Structure THEOCEM* **180**, 79–93 (1988).

70. H. Ihee, A. H. Zewail, and W. A. Goddard, Conformations and barriers of haloethyl radicals (CH_2XCH_2, X = F, Cl, Br, I): *ab initio* studies. *J. Phys. Chem. A* **103**(33), 6638–6649 (1999).

71. M. Rasmusson, A. N. Tarnovsky, T. Pascher, V. Sundstrom, and E. Akesson, Photodissociation of CH_2ICH_2I, CF_2ICF_2I, and CF_2BrCF_2I in solution. *J. Phys. Chem. A* **106**(31), 7090–7098 (2002).

72. A. J. Bowles, A. Hudson, and R. A. Jackson, Hyperfine coupling constants of the 2-chloroethyl and related radicals. *Chem. Phys. Lett.* **5**(9), 552–554 (1970).

73. D. J. Edge and J. K. Kochi, Effects of halogen substitution on alkyl radicals: conformational studies by electron spin resonance. *J. Am. Chem. Soc.* **94**(18), 6485–6495 (1972).

74. S. P. Maj, M. C. R. Symons, and P. M. R. Trousson, Bridged bromine radicals: an electron spin resonance study. *J. Chem. Soc. Chem. Commun.*, 561–562 (1984).

75. J. Fossey, D. Lefort, and J. Sorba, *Free Radicals in Organic Chemistry*, John Wiley & Sons, Hoboken, NJ, 1995.

76. P. S. Skell, D. L. Tuleen, and P. D. Readio, Stereochemical evidence of bridged radicals. *J. Am. Chem. Soc.* **85**(18), 2849–2850 (1963).

77. A. M. Lindenberg, Y. Acremann, D. P. Lowney, P. A. Heimann, T. K. Allison, T. Matthews, and R. W. Falcone, Time-resolved measurements of the structure of water at constant density. *J. Chem. Phys.* **122**(20), 204507 (2005).

78. L. Bosio, S.-H. Chen, and J. Teixeira, Isochoric temperature differential of the X-ray structure factor and structural rearrangements in low-temperature heavy water. *Phys. Rev. A: Gen. Phys.* **27**(3), 1468–1475 (1983).

79. J. A. Polo and P. A. Egelstaff, Neutron-diffraction study of low-temperature water. *Phys. Rev. A: Gen. Phys.* **27**(3), 1508–1514 (1983).

80. J. C. Dore, M. A. M. Sufi, and M. Bellissent-Funel, Structural change in D_2O water as a function of temperature: the isochoric temperature derivative function for neutron diffraction. *Phys. Chem. Chem. Phys.* **2**, 1599–1602 (2000).

81. M. Cammarata, M. Lorenc, T. K. Kim, J. H. Lee, Q. Y. Kong, E. Pontecorvo, M. Lo Russo, G. Schiro, A. Cupane, M. Wulff, and H. Ihee, Impulsive solvent heating probed by picosecond X-ray diffraction. *J. Chem. Phys.* **124**(12), 124504 (2006).

82. U. Kreibig and M. Vollmer, *Optical Properties of Metal Clusters*, Springer Series in Materials Science 2502, Springer, New York, 1995.

83. J. R. Krenn, A. Leitner, and F. R. Aussenegg, Metal nano-optics. In *Encyclopedia of Nanoscience and Nanotechnology*, Vol. 5 (H. S. Nalwa, ed.), American Scientific Publishers, 2004, pp. 411–419.

84. S. Link and M. A. El-Sayed, Shape and size dependence of radiative, non-radiative and photothermal properties of gold nanocrystals. *Int. Rev. Phys. Chem.* **19**(3), 409–453 (2000).

85. A. Plech, V. Kotaidis, S. Gresillon, C. Dahmen, and G. von Plessen, Laser-induced heating and melting of gold nanoparticles studied by time-resolved X-ray scattering. *Phys. Rev. B: Condensed Matter Mater. Phys.* **70**(19), 195423 (2004).

86. V. Kotaidis and A. Plech, Cavitation dynamics on the nanoscale. *Appl. Phys. Lett.* **87**(21), 213102 (2005).

87. A. Plech, V. Kotaidis, M. Lorenc, and J. Boneberg, Femtosecond laser near-field ablation from gold nanoparticles. *Nat. Phys.* **2**, 44–47 (2006).

NONEQUILIBRIUM FLUCTUATIONS IN SMALL SYSTEMS: FROM PHYSICS TO BIOLOGY

FELIX RITORT

Department de Fisica Fonamental, Faculty of Physics, Universitat de Barcelona, Diagonal 647, 08028 Barcelona, Spain

CONTENTS

Advances in Chemical Physics, Volume 137, edited by Stuart A. Rice

I. WHAT ARE SMALL SYSTEMS?

Thermodynamics, a scientific discipline inherited from the 18th century, is facing new challenges in the description of nonequilibrium small (sometimes also called mesoscopic) systems. Thermodynamics is a discipline built in order to explain and interpret energetic processes occurring in macroscopic systems made out of a large number of molecules on the order of the Avogadro number. Although thermodynamics makes general statements beyond reversible processes, its full applicability is found in equilibrium systems where it can make quantitative predictions just based on a few laws. The subsequent development of statistical mechanics has provided a solid probabilistic basis for thermodynamics and increased its predictive power at the same time. The development of statistical mechanics goes together with the establishment of the molecular hypothesis. Matter is made out of interacting molecules in motion. Heat, energy, and work are measurable quantities that depend on the motion of molecules. The laws of thermodynamics operate at all scales.

Let us now consider the case of heat conduction along polymer fibers. Thermodynamics applies at the microscopic or molecular scale, where heat conduction takes place along molecules linked along a single polymer fiber, up to the macroscopic scale where heat is transmitted through all the fibers that make a piece of rubber. The main difference between the two cases is the amount of heat transmitted along the system per unit of time. In the first case the amount of heat can be a few $k_B T$ per millisecond whereas in the second it can be on the order of $N_f k_B T$, where N_f is the number of polymer fibers in the piece of rubber. The relative amplitude of the heat fluctuations are on the order of 1 in the molecular case and $1/\sqrt{N_f}$ in the macroscopic case. Because N_f is usually very large, the relative magnitude of heat fluctuations is negligible for the piece of rubber as compared to the single polymer fiber. We then say that the single polymer fiber is a small system whereas the piece of rubber is a macroscopic system made out of a very large collection of small systems that are assembled together.

Small systems are those in which the energy exchanged with the environment is a few times $k_B T$ and energy fluctuations are observable.

A few can be 10 or 1000 depending on the system. A small system must not necessarily be of molecular size or contain a few numbers of molecules. For example, a single polymer chain may behave as a small system although it contains millions of covalently linked monomer units. At the same time, a molecular system may not be small if the transferred energy is measured over long times compared to the characteristic heat diffusion time. In that case the average energy exchanged with the environment during a time interval t can be as large as desired by choosing t large enough. Conversely, a macroscopic system operating at short time scales could deliver a tiny amount of energy to the environment, small enough for fluctuations to be observable and the system being effectively small.

Because macroscopic systems are collections of many molecules, we expect that the same laws that have been found to be applicable in macroscopic systems are also valid in small systems containing a few numbers of molecules [1, 2]. Yet, the phenomena that we will observe in the two regimes will be different. Fluctuations in large systems are mostly determined by the conditions of the environment. Large deviations from the average behavior are hardly observable and the structural properties of the system cannot be inferred from the spectrum of fluctuations. In contrast, small systems will display large deviations from their average behavior. These turn out to be quite independent of the conditions of the surrounding environment (temperature, pressure, chemical potential) and carry information about the structure of the system and its nonequilibrium behavior. We may then say that information about the *structure* is carried in the tails of the statistical distributions describing molecular properties.

The world surrounding us is mostly out of equilibrium, equilibrium being just an idealization that requires specific conditions to be met in the laboratory. Even today we do not have a general theory about nonequilibrium macroscopic systems as we have for equilibrium ones. Onsager theory is probably the most successful attempt, albeit its domain of validity is restricted to the linear response regime. In small systems the situation seems to be the opposite. Over the past years, a set of theoretical results that go under the name of fluctuation theorems have been unveiled. These theorems make specific predictions about energy processes in small systems that can be scrutinized in the laboratory.

The interest of the scientific community on small systems has been boosted by the recent advent of micromanipulation techniques and nanotechnologies. These provide adequate scientific instruments that can measure tiny energies in physical systems under nonequilibrium conditions. Most of the excitement comes also from the more or less recent observation that biological matter has successfully exploited the smallness of biomolecular structures (such as complexes made out of nucleic acids and proteins) and the fact that they are embedded in a nonequilibrium environment to become wonderfully complex and efficient at the same time [3, 4].

The goal of this chapter is to discuss these ideas from a physicist's perspective by emphasizing the underlying common aspects in a broad category of systems, from glasses to biomolecules. We aim to put together some concepts in statistical mechanics that may become the building blocks underlying a future theory of nonequilibrium small systems. This is not a review in the traditional sense but rather a survey of a few selected topics in nonequilibrium statistical mechanics concerning systems that range from physics to biology. The selection is biased by my own particular taste and expertise. For this reason I have not tried to cover most of the relevant references for each selected topic but rather emphasize a few of them that make explicit connection with my discourse. Interested readers are advised to look at other reviews that have recently been written on related subjects [5–7].

Section II introduces two examples, one from physics and the other from biology, that are paradigms of nonequilibrium behavior. Section III covers most important aspects of fluctuation theorems, whereas Section IV presents applications of fluctuation theorems to physics and biology. Section V presents the discipline of path thermodynamics and briefly discusses large deviation functions. Section VI discusses the topic of glassy dynamics from the perspective of nonequilibrium fluctuations in small cooperatively rearranging regions. We conclude with a brief discussion of future perspectives.

II. SMALL SYSTEMS IN PHYSICS AND BIOLOGY

A. Colloidal Systems

Condensed matter physics is full of examples where nonequilibrium fluctuations of mesoscopic regions govern the nonequilibrium behavior that is observed at the macroscopic level. A class of systems that have attracted a lot of attention for many decades and that still remain poorly understood are glassy systems, such as supercooled liquids and soft materials [8]. Glassy systems can be prepared in a nonequilibrium state (e.g., by fast quenching the sample from high to low temperatures) and subsequently following the time evolution of the system as a function of time (also called age of the system). Glassy systems display extremely slow relaxation and aging behavior, that is, an age-dependent response to the action of an external perturbation. Aging systems respond slower as they *get older*, keeping memory of their age for time scales that range from picoseconds to years. The slow dynamics observed in glassy systems is dominated by intermittent, large, and rare fluctuations, where mesoscopic regions release some stress energy to the environment. Current experimental evidence suggests that these events correspond to structural rearrangements of clusters of molecules inside the glass, which release some energy through an activated and cooperative process. These cooperatively rearranging regions are responsible for the heterogeneous dynamics observed in glassy systems, as these lead to a great disparity of relaxation times. The fact that slow dynamics of glassy systems virtually takes forever indicates that the average amount of energy released in a rearrangement event must be small enough to account for an overall net energy release of the whole sample that is not larger than the stress energy contained in the system in the initial nonequilibrium state.

In some systems, such as colloids, the free volume (i.e., the volume of the system that is available for motion to the colloidal particles) is the relevant variable, and the volume fraction of colloidal particles ϕ is the parameter governing the relaxation rate. Relaxation in colloidal systems is determined by the release of tensional stress energy and free volume in spatial regions that contain a few particles. Colloidal systems offer great advantages to do experiments for several reasons: (1) in colloids the control parameter is the volume fraction, ϕ, a quantity easy to control in experiments; (2) under appropriate solvent conditions colloidal particles behave as hard spheres, a system that is pretty well known and has been theoretically and numerically studied for many years; and (3) the size of colloidal particles is typically a few microns, making it possible to follow the motion of a small number of particles using video microscopy and spectroscopic techniques. This allows one to detect cooperatively rearranging clusters of particles and characterize the heterogeneous dynamics. Experiments have been done with poly(methyl methacrylate) (PMMA) particles of $\simeq 1\,\mu\text{m}$ radius suspended in organic solvents [9, 10]. Confocal microscopy then allows one to acquire images of spatial regions of extension on the order of tens of microns that contain a few thousand of particles, small enough to detect the collective motion of clusters. In experiments carried out by Courtland and Weeks [11], a highly stressed nonequilibrium state is produced by mechanically stirring a colloidal system at volume fractions $\phi \sim \phi_g$, where ϕ_g is the value of the volume fraction at the glass transition where colloidal motion arrests. The subsequent motion is then observed. A few experimental results are shown in Fig. 1. The mean square displacement of the particles inside the confocal region shows aging behavior. Importantly, the region observed is small enough to observe temporal heterogeneity; that is, the aging behavior is not smooth with the age of the system as usually observed in light scattering experiments. Finally, the mean square displacement for a single trajectory shows abrupt events characteristic of collective motions involving a few tens of particles. By analyzing the average number of particles belonging to a single cluster, Courtland and Weeks [11] found that no more than 40 particles participate in the rearrangment of a single cluster, suggesting that cooperatively rearranging regions are not larger than a few particle radii in extension. Large deviations, intermittent events, and heterogeneous kinetics are the main features observed in these experiments.

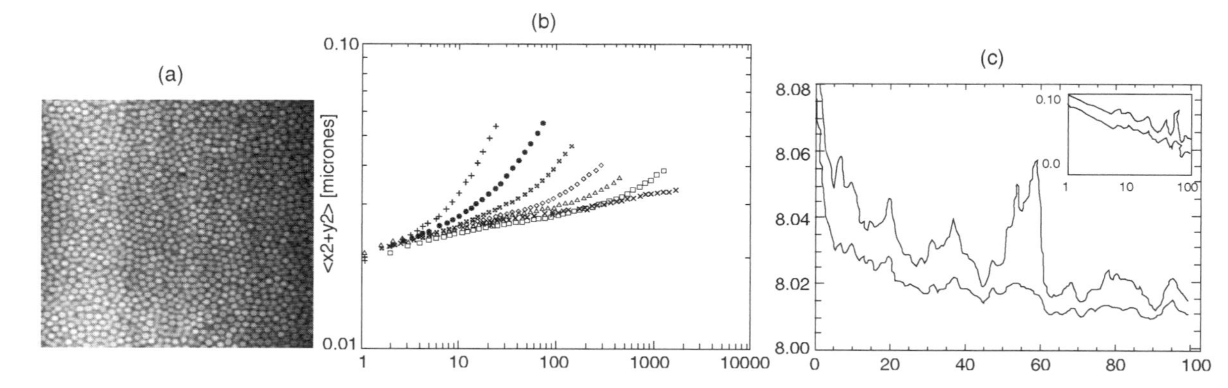

Figure 1. (a) A snapshot picture of a colloidal system obtained with confocal microscope. (b) Aging behavior observed in the mean square displacement, $\langle \Delta x^2 \rangle$, as a function of time for different ages. The colloidal system reorganizes slower as it becomes older. (c) $\gamma = \sqrt{\langle \Delta x^4 \rangle / 3}$ (upper curve) and $\langle \Delta x^2 \rangle$ (lower curve) as a function of the age measured over a fixed time window $\Delta t = 10\,\text{min}$. For a diffusive dynamics both curves should coincide, however these measurements show deviations from diffusive dynamics as well as intermittent behavior. Panels (a) and (b) from http://www.physics.emory.edu/~weeks/lab/aging.html and Panel (c) from Ref. 11. (See color insert.)

B. Molecular Machines

Biochemistry and molecular biology are scientific disciplines aiming to describe the structure, organization, and function of living matter [12, 13]. Both disciplines seek an understanding of life processes in molecular terms. The main objects of study are biological molecules and the function they play in the biological process where they intervene. Biomolecules are small systems from several points of view: first, from their size, where they span just a few nanometers of extension; second, from the energies they require to function properly, which is determined by the amount of energy that can be extracted by hydrolyzing one molecule of ATP (approximately $12k_BT$ at room temperature or 300 K); and third, from the typically short amount of time that it takes to complete an intermediate step in a biological reaction. Inside the cell many reactions that would take an enormous amount of time under nonbiological conditions are speeded up by several orders of magnitude in the presence of specific enzymes.

Molecular machines (also called molecular motors) are amazing complexes made out of several parts or domains that coordinate their behavior to perform specific biological functions by operating out of equilibrium. Molecular machines hydrolyze energy carrier molecules such as ATP to transform the chemical energy contained in the high energy bonds into mechanical motion [14–17]. An example of a molecular machine that has been studied by molecular biologists and biophysicists is the RNA polymerase [18,19]. This is an enzyme that synthesizes an premessenger RNA molecule by translocating along the DNA and reading, step by step, the sequence of bases along the DNA backbone. The readout of the RNA polymerase is exported from the nucleus to the cytoplasm of the cell to later be translated in the ribosome, a huge molecular machine that synthesizes the protein coded into the messenger RNA [20]. By using single molecule experiments, it is possible to grab one DNA molecule by both ends using optical tweezers and follow the translocation motion of the RNA polymerase [21, 22]. Current optical tweezer techniques have even resolved the motion of the enzyme at the level of a single base pair [23, 24]. The experiment requires the flow of enzymes and proteins into the fluidics chamber that are necessary to initiate the transcription reaction. The subsequent motion and transcription by the RNA polymerase is called elongation and can be studied under applied force conditions that assist or oppose the motion of the enzyme [25]. In Fig. 2 we show the results obtained in the Bustamante group for the RNA polymerase of *Escherichia coli*, a bacteria found in the intestinal tracts of animals. In Fig. 2a the polymerase apparently moves at a constant average speed but is characterized by pauses (black arrows) where motion temporarily arrests. In Fig. 2b we show the transcription rate (or speed of the enzyme) as a function of time. Note the large intermittent fluctuations in the transcription

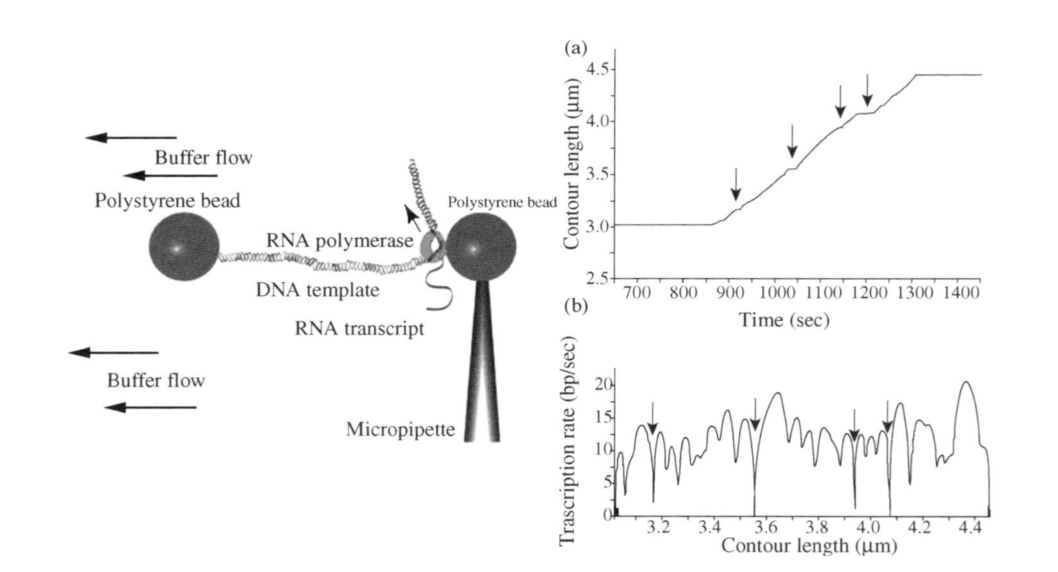

Figure 2. (Left) Experimental setup in force flow measurements. Optical tweezers are used to trap beads but forces are applied on the RNApol–DNA molecular complex using the Stokes drag force acting on the left bead immersed in the flow. In this setup, force assists RNA transcription as the DNA tether between beads increases in length as a function of time. (a) The contour length of the DNA tether as a function of time and (b) the transcription rate as a function of the contour length. Pauses (temporary arrests of transcription) are shown as vertical arrows. (From Ref. 25.) (See color insert.)

rate, a typical feature of small systems embedded in a noisy thermal environment. In contrast to the slow dynamics observed in colloidal systems (Section II.A), the kinetic motion of the polymerase is not progressively slower but steady and fast. We then say that the polymerase is in a nonequilibrium steady state. As in the previous case, large deviations, intermittent events, and complex kinetics are the main features we observe in these experiments.

III. FLUCTUATION THEOREMS

Fluctuation theorems (FTs) make statements about energy exchanges that take place between a system and its surroundings under general nonequilibrium conditions. Since their discovery in the mid -1990s [26–28], there has been an increasing interest to elucidate their importance and implications. FTs provide a fresh new look at old questions such as the origin of irreversibility and the second law in statistical mechanics [29, 30]. In addition, FTs provide statements about energy fluctuations in small systems, which, under generic conditions, should be experimentally observable. FTs have been discussed in the context of deterministic, stochastic and thermostatted systems. Although the results obtained differ depending on the particular model of the dynamics that is used, in a nutshell they are pretty similar.

FTs are related to the so-called nonequilibrium work relations introduced by Jarzynski [31]. This fundamental relation can be seen as a consequence of the FTs [32, 33]. It represents a new result beyond classical thermodynamics that shows the possibility to recover free energy differences using irreversible processes. Several reviews have been written on the subject [3, 34–37] with specific emphasis on theory and/or experiments. In the next sections we review some of the main results. Throughout the text we will take $k_B = 1$.

A. Nonequilibrium States

An important concept in thermodynamics is the state variable. State variables are those that, once determined, uniquely specify the thermodynamic state of the system. Examples are the temperature, the pressure, the volume, and the mass of the different components in a given system. To specify the state variables of a system it is common to put the system in contact with a bath. The bath is any set of sources (of energy, volume, mass, etc.) large enough to remain unaffected by the interaction with the system under study. The bath ensures that a system can reach a given temperature, pressure, volume, and mass concentration of the different components when put in thermal contact with the bath (i.e., with all the relevant sources). Equilibrium states are then generated by putting the system in contact with a bath and waiting until the system properties relax to the equilibrium values. Under such conditions the system properties do not change with time and the average heat/work/mass exchanged between the system and the bath is zero.

Nonequilibrium states can be produced under a great variety of conditions, either by continuously changing the parameters of the bath or by preparing the system in an initial nonequilibrium state that slowly relaxes toward equilibrium. In general, a nonequilibrium state is produced whenever the system properties change with time and/or the net heat/work/mass exchanged by the system and the bath is nonzero. We can distinguish at least three different types of nonequilibrium states:

- **Nonequilibrium Transient State (NETS).** The system is initially prepared in an equilibrium state and later driven out of equilibrium by switching on an external perturbation. The system returns to a new equilibrium state after waiting long enough once the external perturbation stops changing.
- **Nonequilibrium Steady State (NESS).** The system is driven by external forces (either time dependent or nonconservative) in a stationary nonequilibrium state, where its properties do not change with time. The steady state is an irreversible nonequilibrium process that cannot be described by the Boltzmann–Gibbs distribution, where the average heat that is dissipated by the system (equal to the entropy production of the bath) is positive.

  There are still other categories of NESS. For example, in nonequilibrium transient steady states the system starts in a nonequilibrium steady state but is driven out of that steady state by an external perturbation to finally settle in a new steady state.
- **Nonequilibrium Aging State (NEAS).** The system is initially prepared in a nonequilibrium state and put in contact with the sources. The system is then allowed to evolve alone but fails to reach thermal equilibrium in observable or laboratory time scales. In this case the system is in a nonstationary slowly relaxing nonequilibrium state called *aging state* and is characterized by a very small entropy production of the sources. In the aging state two-times correlations decay slower as the system becomes older. Two-time correlation functions depend on both times and not just on their difference.

There are many examples of nonequilibrium states. A classic example of a NESS is an electrical circuit made out of a battery and a resistance. The current flows through the resistance and the chemical energy stored in the battery is dissipated to the environment in the form of heat; the average dissipated power, $\mathcal{P}_{\text{diss}} = VI$, is identical to the power supplied by the battery. Another example is a sheared fluid between two plates or coverslips and one of them is moved relative to the other at a constant velocity v. To sustain such a state, a mechanical power that is equal to $\mathcal{P} \propto \eta v^2$ has to be exerted on the moving plate, where η is the viscosity of water. The mechanical work produced is then dissipated in the form of heat through the viscous friction between contiguous fluid layers. Other examples of the NESS are chemical reactions in metabolic pathways that are sustained by activated carrier molecules such as ATP. In this case, hydrolysis of ATP is strongly coupled to specific oxidative reactions. For example, ionic channels use ATP hydrolysis to transport protons against the electromotive force.

A classic example of a NETS is the case of a protein in its initial native state that is mechanically pulled (e.g., using AFM) by exerting force on the ends of the molecule. The protein is initially folded and in thermal equilibrium with the surrounding aqueous solvent. By mechanically stretching the protein is pulled away from equilibrium into a transient state until it finally settles into the unfolded and extended new equilibrium state. Another example of a NETS is a bead immersed in water and trapped in an optical well generated by a focused laser beam. When the trap is moved to a new position (e.g., by moving the laser beam) the bead is driven into a NETS. After some time the bead again reaches equilibrium at the new position of the trap. In another experiment the trap is suddenly put into motion at a speed v so the bead is transiently driven away from its equilibrium average position until it settles into a NESS characterized by the speed of the trap. This results in the average position of the bead lagging behind the position of the center of the trap.

The classic example of a NEAS is a supercooled liquid cooled below its glass transition temperature. The liquid solidifies into an amorphous, slowly relaxing state characterized by huge relaxational times and anomalous low frequency response. Other systems are colloids that can be prepared in a NEAS by the sudden reduction/increase of the volume fraction of the colloidal particles or by putting the system under a strain/stress.

The classes of nonequilibrium states previously described do not make distinctions based on whether the system is macroscopic or small. In small systems, however, it is common to speak about the control parameter to emphasize the importance of the constraints imposed by the bath that are externally controlled and do not fluctuate. The control parameter (λ) represents a value (in general, a set of values) that defines the state of the bath. Its value determines the equilibrium properties of the system (e.g., the equation of state). In macroscopic systems, it is unnecessary to discern which value is externally controlled because fluctuations are small and all equilibrium ensembles give the same equivalent thermodynamic description (i.e., the same equation of state). Differences arise only when including fluctuations in the description. The nonequilibrium behavior of small systems is then strongly dependent on the protocol used to drive them out of equilibrium. The protocol is generally defined by the time evolution of the control parameter $\lambda(t)$. As a consequence, the characterization of the protocol $\lambda(t)$ is an essential step to unambiguously defining the nonequilibrium state. Figure 3 shows a representation of a few examples of the NESS and control parameters.

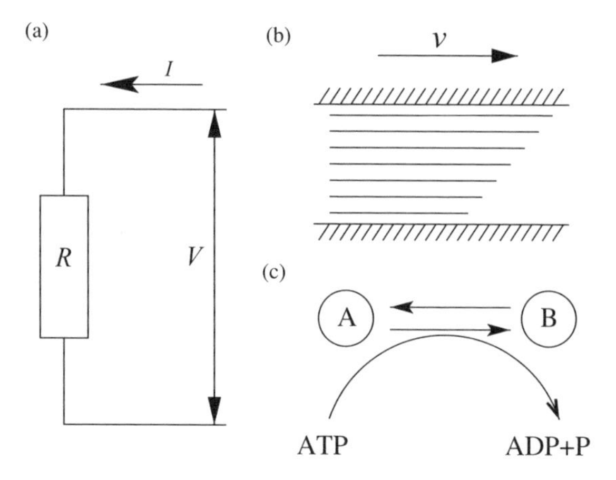

Figure 3. Examples of the NESS. (a) An electric current I flowing through a resistance R and maintained by a voltage source or control parameter V. (b) A fluid sheared between two plates that move at speed v (the control parameter) relative to each other. (c) A chemical reaction $A \to B$ coupled to ATP hydrolysis. The control parameters here are the concentrations of ATP and ADP.

B. Fluctuation Theorems in Stochastic Dynamics

In this section we present a derivation of the FT based on stochastic dynamics. In contrast to deterministic systems, stochastic dynamics naturally incorporates crucial assumptions needed for the derivation, such as the ergodicity hypothesis. The derivation we present here follows the approach introduced by Crooks–Kurchan–Lebowitz–Spohn [38, 39] and includes some results recently obtained by Seifert [40] using Langevin systems.

1. The Master Equation

Let us consider a stochastic system described by a generic variable $\mathcal{C}$. This variable may stand for the position of a bead in an optical trap, the velocity field of a fluid, the current passing through a resistance, of the number of native contacts in a protein. A trajectory or path Γ in configurational space is described by a discrete sequence of configurations in phase space,

$$\Gamma \equiv \{\mathcal{C}_0, \mathcal{C}_1, \mathcal{C}_2, \ldots, \mathcal{C}_M\} \tag{1}$$

where the system occupies configuration $\mathcal{C}_k$ at time $t_k = k\,\Delta t$ and Δt is the duration of the discretized elementary time step. In what follows, we consider paths that start at $\mathcal{C}_0$ at time $t = 0$ and end at the configuration $\mathcal{C}_M$ at time $t = M\,\Delta t$. The continuous time limit is recovered by taking $M \to \infty, \Delta t \to 0$ for a fixed value of t.

Let $\langle(\cdots)\rangle$ denote the average over all paths that start at $t = 0$ at configurations $\mathcal{C}_0$ initially chosen from a distribution $P_0(\mathcal{C})$. We also define $P_k(\mathcal{C})$ as the probability, measured over all possible dynamical paths, that the system is in configuration $\mathcal{C}$ at time $t_k = k\,\Delta t$. Probabilities are normalized for all k,

$$\sum_{\mathcal{C}} P_k(\mathcal{C}) = 1 \tag{2}$$

The system is assumed to be in contact with a thermal bath at temperature T. We also assume that the microscopic dynamics of the system is of the Markovian type: the probability that the system has a given configuration at a given time only depends on its previous configuration. We then introduce the transition probability $\mathcal{W}_k(\mathcal{C} \to \mathcal{C}')$. This denotes the probability for the system to change from $\mathcal{C}$ to $\mathcal{C}'$ at time step k. According to the Bayes formula,

$$P_{k+1}(\mathcal{C}) = \sum_{\mathcal{C}'} \mathcal{W}_k(\mathcal{C}' \to \mathcal{C}) P_k(\mathcal{C}') \tag{3}$$

where the $\mathcal{W}'$ satisfy the normalization condition,

$$\sum_{\mathcal{C}'} \mathcal{W}_k(\mathcal{C} \to \mathcal{C}') = 1 \tag{4}$$

Using Eqs. (2) and (3) we can write the following master equation for the probability $P_k(\mathcal{C})$:

$$\Delta P_k(\mathcal{C}) = P_{k+1}(\mathcal{C}) - P_k(\mathcal{C}) = \sum_{\mathcal{C}' \neq \mathcal{C}} \mathcal{W}_k(\mathcal{C}' \to \mathcal{C}) P_k(\mathcal{C}') - \sum_{\mathcal{C}' \neq \mathcal{C}} \mathcal{W}_k(\mathcal{C} \to \mathcal{C}') P_k(\mathcal{C}) \tag{5}$$

where the terms $\mathcal{C} = \mathcal{C}'$ have not been included as they cancel out in the first and second sums on the right-hand side (rhs). The first term on the rhs accounts for all transitions leading to the configuration $\mathcal{C}$, whereas the second term counts all processes leaving $\mathcal{C}$. It is convenient to introduce the rates $r_t(\mathcal{C} \to \mathcal{C}')$ in the continuous time limit $\Delta t \to 0$,

$$r_t(\mathcal{C} \to \mathcal{C}') = \lim_{\Delta t \to 0} \frac{\mathcal{W}_k(\mathcal{C} \to \mathcal{C}')}{\Delta t}; \quad \forall \mathcal{C} \neq \mathcal{C}' \tag{6}$$

Equation (5) becomes

$$\frac{\partial P_t(\mathcal{C})}{\partial t} = \sum_{\mathcal{C}' \neq \mathcal{C}} r_t(\mathcal{C}' \to \mathcal{C}) P_t(\mathcal{C}') - \sum_{\mathcal{C}' \neq \mathcal{C}} r_r(\mathcal{C} \to \mathcal{C}') P_t(\mathcal{C}) \tag{7}$$

2. *Microscopic Reversibility*

We now introduce the concept of the control parameter λ (see Section III.A). In the present scheme the discrete time sequence $\{\lambda_k; 0 \leq k \leq M\}$ defines the perturbation protocol. The transition probability $\mathcal{W}_k(\mathcal{C} \to \mathcal{C}')$ now depends explicitly on time through the value of an external time-dependent parameter λ_k. The parameter λ_k may indicate any sort of externally controlled variable that determines the state of the system, for instance, the value of the external magnetic field applied on a magnetic system, the value of the mechanical force applied to the ends of a molecule, the position of a piston containing a gas, or the concentrations of ATP and ADP in a molecular reaction coupled to hydrolysis (see Fig. 3). The time variation of the control parameter, $\dot{\lambda} = (\lambda_{k+1} - \lambda_k)/\Delta t$, is used as a tunable parameter, which determines how irreversible the nonequilibrium process is. In order to emphasize the importance of the control parameter, in what follows we will parameterize probabilities and transition probabilities by the value of the control parameter at time step k, λ (rather than by the time t). Therefore, we will write $P_\lambda(\mathcal{C})$ and $\mathcal{W}_\lambda(\mathcal{C} \to \mathcal{C}')$ for the probabilities and transition probabilities, respectively, at a given time t.

The transition probabilities $\mathcal{W}_\lambda(\mathcal{C} \to \mathcal{C}')$ cannot be arbitrary but must guarantee that the equilibrium state $P_\lambda^{\mathrm{eq}}(\mathcal{C})$ is a stationary solution of the master equation (5). The simplest way to impose such a condition is to model the microscopic dynamics as ergodic and reversible for a fixed value of λ:

$$\frac{\mathcal{W}_\lambda(\mathcal{C} \to \mathcal{C}')}{\mathcal{W}_\lambda(\mathcal{C}' \to \mathcal{C})} = \frac{P_\lambda^{\mathrm{eq}}(\mathcal{C}')}{P_\lambda^{\mathrm{eq}}(\mathcal{C})} \tag{8}$$

The latter condition is commonly known as microscopic reversibility or local detailed balance. This property is equivalent to time reversal invariance in deterministic (e.g., thermostatted) dynamics. Although it can be relaxed by requiring just global (rather than detailed) balance, it is physically natural to think of equilibrium as a local property. Microscopic reversibility, a common assumption in nonequilibrium statistical mechanics, is the crucial ingredient in the present derivation.

Equation (8) has been criticized as a relation that is valid only very near to equilibrium because the rates appearing in Eq. (8) are related to the equilibrium distribution $P_\lambda^{\mathrm{eq}}(\mathcal{C})$. However, we must observe that the equilibrium distribution evaluated at a given configuration depends only on the Hamiltonian of the system at that configuration. Therefore, Eq. (8) must be read as a relation that only depends on the energy of configurations, valid close but also far from equilibrium.

Let us now consider all possible dynamical paths Γ that are generated starting from an ensemble of initial configurations at time 0 (described by the initial distribution $P_{\lambda_0}(\mathcal{C})$) and that evolve according to Eq. (8) until time t ($t = M\,\Delta t$, M being the total number of discrete time steps). Dynamical evolution takes place according to a given protocol, $\{\lambda_k, 0 \le k \le M\}$, the protocol defining the nonequilibrium experiment. Different dynamical paths will be generated because of the different initial conditions (weighted with the probability $P_{\lambda_0}(\mathcal{C})$) and because of the stochastic nature of the transitions between configurations at consecutive time steps.

3. *The Nonequilibrium Equality*

Let us consider a generic observable $\mathcal{A}(\Gamma)$. The average value of A is given by

$$\langle \mathcal{A} \rangle = \sum_{\Gamma} P(\Gamma)\mathcal{A}(\Gamma) \tag{9}$$

where Γ denotes the path and $P(\Gamma)$ indicates the probability of that path. Using the fact that the dynamics is Markovian together with the definition Eq. (1), we can write

$$P(\Gamma) = P_{\lambda_0}(\mathcal{C}_0) \prod_{k=0}^{M-1} \mathcal{W}_{\lambda_k}(\mathcal{C}_k \to \mathcal{C}_{k+1}) \tag{10}$$

By inserting Eq. (10) into Eq. (9), we obtain

$$\langle \mathcal{A} \rangle = \sum_{\Gamma} \mathcal{A}(\Gamma) P_{\lambda_0}(\mathcal{C}_0) \prod_{k=0}^{M-1} \mathcal{W}_{\lambda_k}(\mathcal{C}_k \to \mathcal{C}_{k+1}) \tag{11}$$

Using the detailed balance condition Eq. (8), this expression reduces to

$$\langle \mathcal{A} \rangle = \sum_{\Gamma} P_{\lambda_0}(\mathcal{C}_0)\mathcal{A}(\Gamma) \prod_{k=0}^{M-1} \left(\mathcal{W}_{\lambda_k}(\mathcal{C}_{k+1} \to \mathcal{C}_k) \frac{P^{\rm eq}_{\lambda_k}(\mathcal{C}_{k+1})}{P^{\rm eq}_{\lambda_k}(\mathcal{C}_k)} \right) \tag{12}$$

$$= \sum_{\Gamma} \mathcal{A}(\Gamma) P_{\lambda_0}(\mathcal{C}_0) \exp\left(\sum_{k=0}^{M-1} \log\left(\frac{P^{\rm eq}_{\lambda_k}(\mathcal{C}_{k+1})}{P^{\rm eq}_{\lambda_k}(\mathcal{C}_k)} \right) \right) \prod_{k=0}^{M-1} \mathcal{W}_{\lambda_k}(\mathcal{C}_{k+1} \to \mathcal{C}_k) \tag{13}$$

This equation cannot be worked out further. However, let us consider the following observable $\mathcal{S}(\Gamma)$, defined by

$$\mathcal{A}(\Gamma) = \exp(-\mathcal{S}(\Gamma)) = \frac{b(\mathcal{C}_M)}{P_{\lambda_0}(\mathcal{C}_0)} \prod_{k=0}^{M-1} \left(\frac{P^{\rm eq}_{\lambda_k}(\mathcal{C}_k)}{P^{\rm eq}_{\lambda_k}(\mathcal{C}_{k+1})} \right) \tag{14}$$

where $b(\mathcal{C})$ is any positive definite and normalizable function,

$$\sum_{\mathcal{C}} b(\mathcal{C}) = 1 \tag{15}$$

and $P_{\lambda_0}(\mathcal{C}_0) > 0, \forall \mathcal{C}_0$. By inserting Eq. (14) into Eq. (13) we get

$$\langle \exp(-\mathcal{S}) \rangle = \sum_{\Gamma} b(\mathcal{C}_M) \prod_{k=0}^{M-1} \mathcal{W}_{\lambda_k}(\mathcal{C}_{k+1} \to \mathcal{C}_k) = 1 \tag{16}$$

where we have applied a telescopic sum (we first summed over $\mathcal{C}_{M-1}$ by using Eq. (4), and used Eq. (15), and subsequently summed over the rest of variables and used Eq. (4) again). We call $\mathcal{S}(\Gamma)$ the *total dissipation* of the system. It is given by

$$\mathcal{S}(\Gamma) = \sum_{k=0}^{M-1} \log\left(\frac{P^{\text{eq}}_{\lambda_k}(\mathcal{C}_{k+1})}{P^{\text{eq}}_{\lambda_k}(\mathcal{C}_k)}\right) + \log(P_{\lambda_0}(\mathcal{C}_0)) - \log(b(\mathcal{C}_M)) \tag{17}$$

The equality in Eq. (16) immediately implies, by using Jensen's inequality, the following inequality,

$$\langle \mathcal{S} \rangle \geq 0 \tag{18}$$

which is reminiscent of the second law of thermodynamics for nonequilibrium systems: the entropy of the universe (system plus the environment) always increases. Yet, we have to identify the different terms appearing in Eq. (17). It is important to stress that entropy production in nonequilibrium systems can be defined just in terms of the work/heat/mass transferred by the system to the external sources, which represent the bath. The definition of the total dissipation in Eq. (17) is arbitrary because it depends on an undetermined function $b(\mathcal{C})$, Eq. (15). Therefore, the total dissipation S may not necessarily have a general physical meaning and could be interpreted in different ways depending on the specific nonequilibrium context.

Equation (16) has appeared in the past in the literature [41, 42] and is mathematically identical to the Jarzynski equality [31]. We analyze this connection in Section III.C.1.

4. *The Fluctuation Theorem*

A physical insight on the meaning of the total dissipation $\mathcal{S}$ can be obtained by deriving the fluctuation theorem. We start by defining the reverse path Γ^* of a given path Γ. Let us consider the path $\Gamma \equiv \mathcal{C}_0 \to \mathcal{C}_1 \to \cdots \to \mathcal{C}_M$ corresponding to the forward (F) protocol, which is described by the sequence of values of λ at different time steps k, λ_k. Every transition occurring at time step k , $\mathcal{C}_k \rightarrow \mathcal{C}_{k+1}$, is governed by the transition probability $\mathcal{W}_{\lambda_k}(\mathcal{C}_k \rightarrow \mathcal{C}_{k+1})$. The reverse path of Γ is defined as the time reverse sequence of configurations, $\Gamma^* \equiv \mathcal{C}_M \rightarrow \mathcal{C}_{M-1} \rightarrow \cdots \rightarrow \mathcal{C}_0$ corresponding to the reverse (R) protocol described by the time-reversed sequence of values of λ, $\lambda_k^{\mathrm{R}} = \lambda_{M-k-1}$.

The probabilities of a given path and its reverse are given by

$$\mathcal{P}_{\mathrm{F}}(\Gamma) = \prod_{k=0}^{M-1} \mathcal{W}_{\lambda_k}(\mathcal{C}_k \rightarrow \mathcal{C}_{k+1}) \tag{19}$$

$$\mathcal{P}_{\mathrm{R}}(\Gamma^*) = \prod_{k=0}^{M-1} \mathcal{W}_{\lambda_k^{\mathrm{R}}}(\mathcal{C}_{M-k} \rightarrow \mathcal{C}_{M-k-1}) = \prod_{k=0}^{M-1} \mathcal{W}_{\lambda_k}(\mathcal{C}_{k+1} \rightarrow \mathcal{C}_k) \tag{20}$$

where in the last line we shifted variables $k \rightarrow M - 1 - k$. We use the notation $\mathcal{P}$ for the path probabilities rather than the usual letter P. This difference in notation is introduced to stress the fact that path probabilities (Eqs. (19) and (20)) are nonnormalized conditional probabilities; that is, $\sum_\Gamma \mathcal{P}_{\mathrm{F(R)}}(\Gamma) \neq 1$. By using Eq. (8) we get

$$\frac{\mathcal{P}_{\mathrm{F}}(\Gamma)}{\mathcal{P}_{\mathrm{R}}(\Gamma^*)} = \prod_{k=0}^{M-1} \frac{P_{\lambda_k}^{\mathrm{eq}}(\mathcal{C}_{k+1})}{P_{\lambda_k}^{\mathrm{eq}}(\mathcal{C}_k)} = \exp(S_{\mathrm{p}}(\Gamma)) \tag{21}$$

where we defined the *entropy production* of the system,

$$S_{\mathrm{p}}(\Gamma) = \sum_{k=0}^{M-1} \log\left(\frac{P_{\lambda_k}^{\mathrm{eq}}(\mathcal{C}_{k+1})}{P_{\lambda_k}^{\mathrm{eq}}(\mathcal{C}_k)}\right) \tag{22}$$

Note that $S_p(\Gamma)$ is just a part of the total dissipation introduced in Eq. (17),

$$\mathcal{S}(\Gamma) = S_{\mathrm{p}}(\Gamma) + B(\Gamma) \tag{23}$$

where $B(\Gamma)$ is the *boundary term*,

$$B(\Gamma) = \log(P_{\lambda_0}(\mathcal{C}_0)) - \log(b(\mathcal{C}_M)) \tag{24}$$

We tend to identify $S_{\mathrm{p}}(\Gamma)$ as the entropy production in a nonequilibrium system, whereas $B(\Gamma)$ is a term that contributes just at the beginning and end of the nonequilibrium process. Note that the entropy production $S_{\mathrm{p}}(\Gamma)$ is antisymmetric under time reversal, $S_{\mathrm{p}}(\Gamma^*) = -S_{\mathrm{p}}(\Gamma)$, expressing the fact that the entropy production is a quantity related to irreversible motion. According to Eq. (21) paths that produce a given amount of entropy are much more probable than those that consume the same amount of entropy. How improbable entropy consumption is depends exponentially on the amount of entropy consumed. The larger the system is, the larger the probability to produce (rather than consume) a given amount of entropy $S_{\rm p}$.

Equation (21) already has the form of a fluctuation theorem. However, in order to get a proper fluctuation theorem we need to specify relations between probabilities for physically measurable observables rather than paths. From Eq. (21) it is straightforward to derive a fluctuation theorem for the total dissipation $\mathcal{S}$. Let us take $b(\mathcal{C}) = P_{\lambda_M}(\mathcal{C})$. With this choice we get

$$\begin{aligned} \mathcal{S}(\Gamma) = S_{\rm p}(\Gamma) + B(\Gamma) &= \sum_{k=0}^{M-1} \log\left(\frac{P^{\rm eq}_{\lambda_k}(\mathcal{C}_{k+1})}{P^{\rm eq}_{\lambda_k}(\mathcal{C}_k)}\right) \\ &+ \log(P_{\lambda_0}(\mathcal{C}_0)) - \log(P_{\lambda_M}(\mathcal{C}_M)) \end{aligned} \tag{25}$$

The physical motivation behind this choice is that $\mathcal{S}$ now becomes an antisymmetric observable under time reversal. Albeit $S_{\rm p}(\Gamma)$ is always antisymmetric, the choice of Eq. (25) is the only one that guarantees that the total dissipation $\mathcal{S}$ changes sign upon reversal of the path, $\mathcal{S}(\Gamma^*) = -\mathcal{S}(\Gamma)$. The symmetry property of observables under time reversal and the possibility of considering boundary terms where $\mathcal{S}$ is symmetric (rather than antisymmetric) under time reversal has been discussed in Ref. 43.

The probability of producing a total dissipation $\mathcal{S}$ along the forward protocol is given by

$$\begin{aligned} P_{\rm F}(\mathcal{S}) &= \sum_{\Gamma} P_{\lambda_0}(\mathcal{C}_0)\mathcal{P}_{\rm F}(\Gamma)\delta(\mathcal{S}(\Gamma) - \mathcal{S}) \\ &= \sum_{\Gamma} P_{\lambda_0}(\mathcal{C}_0)\mathcal{P}_{\rm R}(\Gamma^*)\exp(S_{\rm p}(\Gamma))\delta(\mathcal{S}(\Gamma) - \mathcal{S}) \\ &= \sum_{\Gamma} P_{\lambda_M}(\mathcal{C}_M)\mathcal{P}_{\rm R}(\Gamma^*)\exp(\mathcal{S}(\Gamma))\delta(\mathcal{S}(\Gamma) - \mathcal{S}) \\ &= \exp(\mathcal{S})\sum_{\Gamma^*} P_{\lambda_M}(\mathcal{C}_M)\mathcal{P}_{\rm R}(\Gamma^*)\delta(\mathcal{S}(\Gamma^*) + \mathcal{S}) = \exp(\mathcal{S})P_{\rm R}(-\mathcal{S}) \end{aligned} \tag{26}$$

In the first line of the derivation we used Eq. (21), in the second we used Eq. (25), and in the last line we took into account the antisymmetric property of $\mathcal{S}(\Gamma)$ and the unicity of the assignment $\Gamma \to \Gamma^*$. This result is known under the generic name of *fluctuation theorem*,

$$\frac{P_{\rm F}(\mathcal{S})}{P_{\rm R}(-\mathcal{S})} = \exp(\mathcal{S}) \tag{27}$$

It is interesting to observe that this relation is not satisfied by the entropy production because the inclusion of a boundary term, Eq. (24), in the total dissipation is required to respect the fluctuation symmetry. In what follows we discuss some of its consequences in some specific situations.

- **Jarzynski Equality.** The nonequilibrium equality, Eq. (16), is just a consequence of Eq. (27) that is obtained by rewriting it as $P_R(-\mathcal{S}) = P_F(\mathcal{S})\exp(-\mathcal{S})$ and integrating both sides of the equation from $\mathcal{S} = -\infty$ to $\mathcal{S} = \infty$.
- **Linear Response Regime.** Equation (27) is trivially satisfied for $\mathcal{S} = 0$ if $P_F(0) = P_R(0)$. The process where $P_{F(R)}(\mathcal{S}) = \delta(\mathcal{S})$ is called quasistatic or reversible. When $\mathcal{S}$ is different from zero but small ($\mathcal{S} < 1$), we can expand Eq. (27) around $\mathcal{S} = 0$ to obtain

$$\begin{aligned} \mathcal{S}P_F(\mathcal{S}) &= \mathcal{S}\exp(\mathcal{S})P_R(-\mathcal{S}) \\ \langle \mathcal{S} \rangle_F &= \langle (-\mathcal{S} + \mathcal{S}^2) \rangle_R + \mathcal{O}(\mathcal{S}^3) \\ \langle (\mathcal{S}^2) \rangle_{F(R)} &= 2\langle \mathcal{S} \rangle_{F(R)} \end{aligned} \tag{28}$$

where we used $\langle \mathcal{S} \rangle_F = \langle \mathcal{S} \rangle_R$, valid up to second order in $\mathcal{S}$. Note the presence of the subindex F(R) for the expectation values in the last line of Eq. (28), which emphasizes the equality of these averages along the forward and reverse processes. Equation (28) is a version of the fluctuation-dissipation theorem (FDT) valid in the linear response region and equivalent to the Onsager reciprocity relations [44].

C. Applications of the FT to Nonequilibrium States

The FT in Eq. (27) finds application in several nonequilibrium contexts. Here we describe specific results for transient and steady states.

1. Nonequilibrium Transient States (NETSs)

We will assume a system initially in thermal equilibrium that is transiently brought to a nonequilibrium state. We are going to show that, under such conditions, the entropy production in Eq. (22) is equal to the heat delivered by the system to the sources. We rewrite Eq. (22) by introducing the potential energy function $G_\lambda(\mathcal{C})$,

$$P_\lambda^{\mathrm{eq}}(\mathcal{C}) = \frac{\exp(-G_\lambda(\mathcal{C}))}{\mathcal{Z}_\lambda} = \exp(-G_\lambda(\mathcal{C}) + \mathcal{G}_\lambda) \tag{29}$$

where $\mathcal{Z}_\lambda = \sum_{\mathcal{C}} \exp(-G_\lambda(\mathcal{C})) = \exp(-\mathcal{G}_\lambda)$ is the partition function and $\mathcal{G}_\lambda$ is the thermodynamic potential. The existence of the potential $G_\lambda(\mathcal{C})$ and the thermodynamic potential $\mathcal{G}_\lambda$ is guaranteed by the Boltzmann–Gibbs ensemble theory. For simplicity we will consider here the canonical ensemble, where the volume V, the number of particles N, and the temperature T are fixed. Needless to say, the following results can be generalized to arbitrary ensembles. In the canonical case $G_\lambda(\mathcal{C})$ is equal to $E_\lambda(\mathcal{C})/T$, where $E_\lambda(\mathcal{C})$ is the total energy function (that includes the kinetic plus the potential terms). $\mathcal{G}_\lambda$ is equal to $F_\lambda(V,T,N)/T$, where F_λ stands for the Helmholtz free energy.

With these definitions the entropy production in Eq. (22) is given by

$$S_{\rm p}(\Gamma) = \sum_{k=0}^{M-1}(G_{\lambda_k}(\mathcal{C}_k) - G_{\lambda_k}(\mathcal{C}_{k+1})) = \frac{1}{T}\sum_{k=0}^{M-1}(E_{\lambda_k}(\mathcal{C}_k) - E_{\lambda_k}(\mathcal{C}_{k+1})) \qquad (30)$$

For the boundary term, Eq. (24), let us take $b(\mathcal{C}) = P^{\rm eq}_{\lambda_M}(\mathcal{C})$:

$$\begin{aligned} B(\Gamma) &= \log(P^{\rm eq}_{\lambda_0}(\mathcal{C}_0)) - \log(P^{\rm eq}_{\lambda_M}(\mathcal{C}_M)) \\ &= G_{\lambda_M}(\mathcal{C}_M) - G_{\lambda_0}(\mathcal{C}_0) - \mathcal{G}_{\lambda_M} + \mathcal{G}_{\lambda_0} \\ &= \frac{1}{T}(E_{\lambda_M}(\mathcal{C}_M) - E_{\lambda_0}(\mathcal{C}_0) - F_{\lambda_M} + F_{\lambda_0}) \end{aligned} \qquad (31)$$

The total dissipation, Eq. (25), is then equal to

$$\mathcal{S}(\Gamma) = S_{\rm p}(\Gamma) + \frac{1}{T}(E_{\lambda_M}(\mathcal{C}_M) - E_{\lambda_0}(\mathcal{C}_0) - F_{\lambda_M} + F_{\lambda_0}) \qquad (32)$$

which can be rewritten as a balance equation for the variation of the energy $E_\lambda(\mathcal{C})$ along a given path,

$$\Delta E(\Gamma) = E_{\lambda_M}(\mathcal{C}_M) - E_{\lambda_0}(\mathcal{C}_0) = T\mathcal{S}(\Gamma) + \Delta F - TS_{\rm p}(\Gamma) \qquad (33)$$

where $\Delta F = F_{\lambda_M} - F_{\lambda_0}$. This is the first law of thermodynamics, where we have identified the term on the left-hand side (lhs) with the total variation of the internal energy $\Delta E(\Gamma)$. Whereas $T\mathcal{S}(\Gamma) + \Delta F$ and $TS_{\rm p}(\Gamma)$ are identified with the mechanical work exerted on the system and the heat delivered to the bath, respectively,

$$\Delta E(\Gamma) = W(\Gamma) - Q(\Gamma) \qquad (34)$$

$$W(\Gamma) = T\mathcal{S}(\Gamma) + \Delta F \qquad (35)$$

$$Q(\Gamma) = TS_{\rm p}(\Gamma) \qquad (36)$$

By using Eq. (30) we obtain the following expressions for work and heat:

$$W(\Gamma) = \sum_{k=0}^{M-1}(E_{\lambda_{k+1}}(\mathcal{C}_{k+1}) - E_{\lambda_k}(\mathcal{C}_{k+1})) \qquad (37)$$

$$Q(\Gamma) = \sum_{k=0}^{M-1}(E_{\lambda_k}(\mathcal{C}_k) - E_{\lambda_k}(\mathcal{C}_{k+1})) \qquad (38)$$

The physical meaning of both entropies is now clear. Whereas $S_{\rm p}$ stands for the heat transferred by the system to the sources (Eq. (36)), the total dissipation term $T\mathcal{S}$ (Eq. (35)) is just the difference between the total mechanical work exerted on the system, $W(\Gamma)$, and the reversible work, $W_{\rm rev} = \Delta F$. It is customary to define this quantity as the dissipated work, $W_{\rm diss}$:

$$W_{\rm diss}(\Gamma) = T\mathcal{S}(\Gamma) = W(\Gamma) - \Delta F = W(\Gamma) - W_{\rm rev} \tag{39}$$

The nonequilibrium equality in Eq. (16) becomes the nonequilibrium work relation originally derived by Jarzynski using Hamiltonian dynamics [31],

$$\langle \exp(-W_{\rm diss}/T) \rangle = 1 \quad \text{or} \quad \langle \exp(-W/T) \rangle = \exp(-\Delta F/T) \tag{40}$$

This relation is called the *Jarzynski equality* (hereafter referred to as JE) and can be used to recover free energies from nonequilibrium simulations or experiments (see Section IV.B.2). The FT in Eq. (27) becomes the *Crooks fluctuation theorem* (hereafter referred to as CFT) [45, 46]:

$$\frac{P_{\rm F}(W_{\rm diss})}{P_{\rm R}(-W_{\rm diss})} = \exp\left(\frac{W_{\rm diss}}{T}\right) \quad \text{or} \quad \frac{P_{\rm F}(W)}{P_{\rm R}(-W)} = \exp\left(\frac{W - \Delta F}{T}\right) \tag{41}$$

The second law of thermodynamics $\overline{W} \geq \Delta F$ also follows naturally as a particular case of Eq. (18) by using Eqs. (39) and (40). Note that for the heat Q a relation equivalent to Eq. (41) does not exist. We mention three aspects of the JE and the CFT.

- **The Fluctuation-Dissipation Parameter *R*.** In the limit of small dissipation $W_{\rm diss} \to 0$, the linear response result, Eq. (28), holds. It is then possible to introduce a parameter R that measures deviations from the linear response behavior.[1] It is defined as

$$R = \frac{\sigma_W^2}{2TW_{\rm diss}} \tag{42}$$

where $\sigma_W^2 = \langle W^2 \rangle - \langle W \rangle^2$ is the variance of the work distribution. In the limit $W_{\rm diss} \to 0$, a second-order cumulant expansion in Eq. (40) gives that R is equal to 1 and Eq. (28) holds. Deviations from $R = 1$ are often interpreted as deviations of the work distribution from a Gaussian. When

[1] Sometimes R is called the fluctuation-dissipation ratio, not to be confused with the identically called but different quantity introduced in glassy systems (see Section VI.B) that quantifies deviations from the fluctuation-dissipation theorem that is valid in equilibrium.

the work distribution is nonGaussian, the system is far from the linear response regime and Eq. (28) is not satisfied anymore.

- **The Kirkwood Formula.** A particular case of the JE, Eq. (40), is the Kirkwood formula [47, 48]. It corresponds to the case where the control parameter only takes two values λ_0 and λ_1. The system is initially in equilibrium at the value λ_0 and, at an arbitrary later time t, the value of λ instantaneously switches to λ_1. In this case Eq. (37) reads

$$W(\Gamma) = \Delta E_\lambda(\mathcal{C}) = E_{\lambda_1}(\mathcal{C}) - E_{\lambda_0}(\mathcal{C}) \tag{43}$$

In this case a path corresponds to a single configuration, $\Gamma \equiv \mathcal{C}$, and Eq. (40) becomes

$$\overline{\exp\left(-\frac{\Delta E_\lambda(\mathcal{C})}{T}\right)} = \exp\left(-\frac{\Delta F}{T}\right) \tag{44}$$

the average $\overline{(\cdots)}$ is taken over all configurations $\mathcal{C}$ sampled according to the equilibrium distribution taken at λ_0, $P^{\text{eq}}_{\lambda_0}(\mathcal{C})$.

- **Heat Exchange Between Two Bodies.** Suppose that we take two bodies initially at equilibrium at temperatures T_{H} and T_{C}, where T_{H} and T_{C} stand for a hot and a cold temperature, respectively. At time $t = 0$ we put them in contact and ask about the probability distribution of heat flow between them. In this case, no work is done between the two bodies and the heat transferred is equal to the energy variation of each of the bodies. Let Q be equal to the heat transferred from the hot to the cold body in one experiment. It can be shown [49] that in this case the total dissipation $\mathcal{S}$ is given by

$$\mathcal{S} = Q\left(\frac{1}{T_{\text{C}}} - \frac{1}{T_{\text{H}}}\right) \tag{45}$$

and the equality in Eq. (40) reads

$$\left\langle \exp\left(-Q\left(\frac{1}{T_{\text{C}}} - \frac{1}{T_{\text{H}}}\right)\right) = 1\right\rangle \tag{46}$$

showing that, on average, net heat is always transferred from the hot to the cold body. Yet, sometimes, we also expect some heat to flow from the cold to the hot body. Again, the probability of such events will be exponentially small with the size of the system.

2. *Nonequilibrium Steady States (NESSs)*

Most investigations on nonequilibrium systems were initially carried out in the NESS. It is widely believed that NESSs are among the best candidate nonequilibrium systems to possibly extend the Boltzmann–Gibbs ensemble theory beyond equilibrium [50, 51].

We can distinguish two types of NESS: time-dependent conservative (C) systems and nonconservative (NC) systems. In the C case the system is acted by a time-dependent force that derives from an external potential. In the NC case the system is driven by (time dependent or not) nonconservative forces. In C systems the control parameter λ has the usual meaning: it specifies the set of external parameters that, once fixed, determine an equilibrium state. Examples are a magnetic dipole in an oscillating field (λ is the value of the time-dependent magnetic field), a bead confined on a moving optical trap and dragged through water (λ is the position of the center of the moving trap), and a fluid sheared between two plates (λ is the time-dependent relative position of the upper and lower plates). In C systems we assume local detailed balance so Eq. (8) still holds.

In contrast to the C case, in NC systems the local detailed balance property, in the form of Eq. (8), does not hold because the system reaches not thermal equilibrium but a stationary or steady state. It is then customary to characterize the NESS by the parameter λ and the stationary distribution by $P_\lambda^{\text{ss}}(\mathcal{C})$. NESS systems in the linear regime (i.e., not driven arbitrarily far from equilibrium) satisfy the Onsager reciprocity relations, where the fluxes are proportional to the forces. The NESS can be maintained by keeping constant either the forces or the fluxes. Examples of NC systems are the flow of a current in an electric circuit (e.g., $\lambda = I, \Delta V$ is either the constant current flowing through the circuit or the constant voltage difference), a Poiseuille fluid flow inside a cylinder (λ could be either the constant fluid flux, Φ, or the pressure difference, ΔP), heat flowing between two sources kept at two different temperatures (λ could be either the heat flux, J_Q, or the temperature difference, ΔT), and the particle exclusion process ($\lambda = \mu^{+,-}$ are the rates of inserting and removing particles at both ends of the chain). In the NESS of NC type, the local detailed balance property of Eq. (8) holds but we replace $P_\lambda^{\text{eq}}(\mathcal{C})$ by the corresponding stationary distribution, $P_\lambda^{\text{ss}}(\mathcal{C})$:

$$\frac{\mathcal{W}_\lambda(\mathcal{C} \to \mathcal{C}')}{\mathcal{W}_\lambda(\mathcal{C}' \to \mathcal{C})} = \frac{P_\lambda^{\text{ss}}(\mathcal{C}')}{P_\lambda^{\text{ss}}(\mathcal{C})} \tag{47}$$

In a steady state in an NC system λ is maintained constant. Because the local form of detailed balance, Eq. (47), holds, the main results of Section III follow. In particular, the nonequilibrium equality in Eq. (16) and the FT in Eq. (27) are still true. However, there is an important difference. In steady states the reverse process is identical to the forward process, $P_{\text{F}}(\mathcal{S}) = P_{\text{R}}(\mathcal{S})$, because λ is maintained constant. Therefore, Eq. (16) and Eq. (27) become

$$\langle \exp(-\mathcal{S}) \rangle = 1 \tag{48}$$

$$\frac{P(\mathcal{S})}{P(-\mathcal{S})} = \exp(\mathcal{S}) \tag{49}$$

We can now extract a general FT for the entropy production S_{p} in the NESS. Let us assume that, on average, S_{p} grows linearly with time, that is, $S_{\text{p}} \gg B$ for large t.

Because $\mathcal{S} = S_{\text{p}} + B$ (Eq. (23)), in the large t limit fluctuations in $\mathcal{S}$ are asymptotically dominated by fluctuations in S_{p}. On average, fluctuations in S_{p} grow like $\sqrt{t}$, whereas fluctuations in the boundary term are finite.

Therefore, Eq. (23) should be asymptotically valid in the large t limit. By taking the logarithm of the right expression we obtain

$$S = \log(P(S)) - \log(P(-S)) \rightarrow S_{\text{p}} + B = \log(P(S_{\text{p}} + B)) - \log(P(S_{\text{p}} - B)) \quad (50)$$

In the NESS, the entropy produced, $S_{\text{p}}(\Gamma)$, along paths of duration t is a fluctuating quantity. By expanding Eq. (50) around S_{p}, we get

$$S_{\text{p}} = \log\left(\frac{P(S_{\text{p}})}{P(-S_{\text{p}})}\right) + B\left(\frac{P'(S_{\text{p}})}{P(S_{\text{p}})} + \frac{P'(-S_{\text{p}})}{P(-S_{\text{p}})} - 1\right) \quad (51)$$

The average entropy production $\langle S_{\text{p}} \rangle$ is defined by averaging S_{p} along an infinite number of paths. Dividing Eq. (51) by $\langle S_{\text{p}} \rangle$ we get

$$\frac{S_{\text{p}}}{\langle S_{\text{p}} \rangle} = \frac{1}{\langle S_{\text{p}} \rangle} \log\left(\frac{P(S_{\text{p}})}{P(-S_{\text{p}})}\right) + \frac{B}{\langle S_{\text{p}} \rangle}\left(\frac{P'(S_{\text{p}})}{P(S_{\text{p}})} + \frac{P'(-S_{\text{p}})}{P(-S_{\text{p}})} - 1\right) \quad (52)$$

We introduce a quantity a that is equal to the ratio between the entropy production and its average value, $a = S_{\text{p}}/\langle S_{\text{p}} \rangle$. We can define the function

$$f_t(a) = \frac{1}{\langle S_{\text{p}} \rangle} \log\left(\frac{P(a)}{P(-a)}\right) \quad (53)$$

Equation (52) can be rewritten as

$$f_t(a) = a - \frac{B}{\langle S_{\text{p}} \rangle}\left(\frac{P'(S_{\text{p}})}{P(S_{\text{p}})} + \frac{P'(-S_{\text{p}})}{P(-S_{\text{p}})} - 1\right) \quad (54)$$

In the large time limit, assuming that $\log(P(S_{\text{p}})) \approx t$, and because B is finite, the second term vanishes relative to the first and $f_t(a) = a + \mathcal{O}(1/t)$. Substituting this result into Eq. (53) we find that an FT holds in the large t limit. However, this is not necessarily, always true. Even for very large t there can be strong deviations in the initial and final states that can make the boundary term B large enough to be comparable to $\langle S_{\text{p}} \rangle$. In other words, for certain initial and/or final conditions, the second term on the rhs of Eq. (54) can be on the same order and comparable to the first term, a. The boundary term can be neglected only if we restrict the size of such large deviations; that is, if we require $|a| \leq a^*$, where a^* is a maximum given value. With this proviso, the FT in a NESS reads

$$\lim_{t\to\infty} \frac{1}{\langle S_{\text{p}} \rangle} \log\left(\frac{P(a)}{P(-a)}\right) = a; \quad |a| \leq a^* \quad (55)$$

In general, it can be very difficult to determine the nature of the boundary terms. A specific result in an exactly solvable case is discussed in Section IV.A.2. Equation (55) is the Gallavotti–Cohen FT derived in the context of deterministic Anosov systems [28]. In that case, S_p stands for the so-called phase space compression factor. It has been experimentally tested by Ciliberto and co-workers in Rayleigh–Bernard convection [52] and turbulent flows [53]. Similar relations have also been tested in athermal systems, for example, in fluidized granular media [54] or the case of two-level systems in fluorescent diamond defects excited by light [55].

The FT in Eq. (27) also describes fluctuations in the total dissipation for transitions between steady states, where λ varies according to a given protocol. In that case, the system starts at time 0 in a given steady state, $P^{ss}_{\lambda_0}(\mathcal{C})$, and evolves away from that steady state at subsequent times. The boundary term for steady-state transitions is then given by

$$B(\Gamma) = \log(P^{ss}_{\lambda_0}(\mathcal{C}_0)) - \log(P^{ss}_{\lambda_M}(\mathcal{C}_M)) \tag{56}$$

where we have chosen the boundary function $b(\mathcal{C}) = P^{ss}_{\lambda_M}(\mathcal{C})$. In that case, the total dissipation is antisymmetric under the time-reversal operation and Eq. (27) holds. Only in cases where the reverse process is equivalent to the forward process is Eq. (49) an exact result. Transitions between nonequilibrium steady states and definitions of the function S have been considered by Hatano and Sasa [56] in the context of Langevin systems.

IV. EXAMPLES AND APPLICATIONS

In this section we analyze in detail two cases where analytical calculations are available and FTs have been experimentally tested: one extracted from physics, the other from biology. We first analyze the bead in a trap and later consider single molecule pulling experiments. These examples show that there are lots of interesting observations that can be made by comparing theory and nonequilibrium experiments in simple systems.

A. A Physical System: A Bead in an Optical Trap

It is very instructive to work out in detail the fluctuations of a bead trapped in a moving potential. This case is of great interest for at least two reasons. First, it provides a simple example of both a NETS and a NESS that can be analytically solved in detail. Second, it can be experimentally realized by trapping micron-sized beads using optical tweezers. The first experiments studying nonequilibrium fluctuations in a bead in a trap were carried out by Evans and collaborators [57] and later on extended in a series of works [58, 59]. Mazonka and Jarzynski [60] and later Van Zon and Cohen [61–63] have carried out detailed theoretical calculations of heat and work fluctuations. Recent experiments have also analyzed the case of a particle in a nonharmonic optical potential [64]. These results have greatly clarified the general validity of the FT and the role of the boundary terms appearing in the total dissipation $\mathcal{S}$.

The case of a bead in a trap is also equivalent to the power fluctuations in a resistance in an *RC* electrical circuit [65] (see Fig. 4). The experimental setup is shown in Fig. 5. A micron-sized bead is immersed in water and trapped in an optical well. In the simplest case the trapping potential is harmonic. Here we will assume that the potential well can have an arbitrary shape and carry out specific analytical computations for the harmonic case.

Let x be the position of the bead in the laboratory frame and $U(x-x^*)$ be the trapping potential of a laser focus that is centered at a reference position x^*. For harmonic potentials we will take $U(x)=\frac{1}{2}\kappa x^2$. By changing the value of x^* the trap is shifted along the x coordinate. A nonequilibrium state can be generated by changing the value of x^* according to a protocol $x^*(t)$. In the notation of the previous sections, $\lambda\equiv x^*$ is the control parameter and $\mathcal{C}\equiv x$ is the configuration. A path Γ starts at $x(0)$ at time 0 and ends at $x(t)$ at time t, $\Gamma\equiv\{x(s);0\leq s\leq t\}$.

At low Reynolds number the motion of the bead can be described by a one-dimensional Langevin equation that contains only the overdamping term,

$$\gamma\dot{x}=f_{x^*}(x)+\eta;\quad \langle\eta(t)\eta(s)\rangle=2T\gamma\delta(t-s) \tag{57}$$

where x is the position of the bead in the laboratory frame, γ is the friction coefficient, $f_{x^*}(x)$ is a conservative force deriving from the trap potential $U(x-x^*)$,

$$f_{x^*}(x)=-(U(x-x^*))'=-\left(\frac{\partial U(x-x^*)}{\partial x}\right) \tag{58}$$

and η is a stochastic white noise.

In equilibrium $x^*(t)=x^*$ is constant in time. In this case, the stationary solution of the master equation is the equilibrium solution

$$P^{\text{eq}}_{x^*}(x)=\frac{\exp\left(-\frac{U(x-x^*)}{T}\right)}{\int dx\exp(-\beta U(x-x^*))}=\frac{\exp\left(-\frac{U(x-x^*)}{T}\right)}{\mathcal{Z}} \tag{59}$$

where $\mathcal{Z}=\int dx\exp(-U(x)/T)$ is the partition function that is independent of the reference position x^*. Because the free energy $F=-T\log(\mathcal{Z})$ does not depend on the control parameter x^*, the free energy change is always zero for arbitrary translations of the trap.

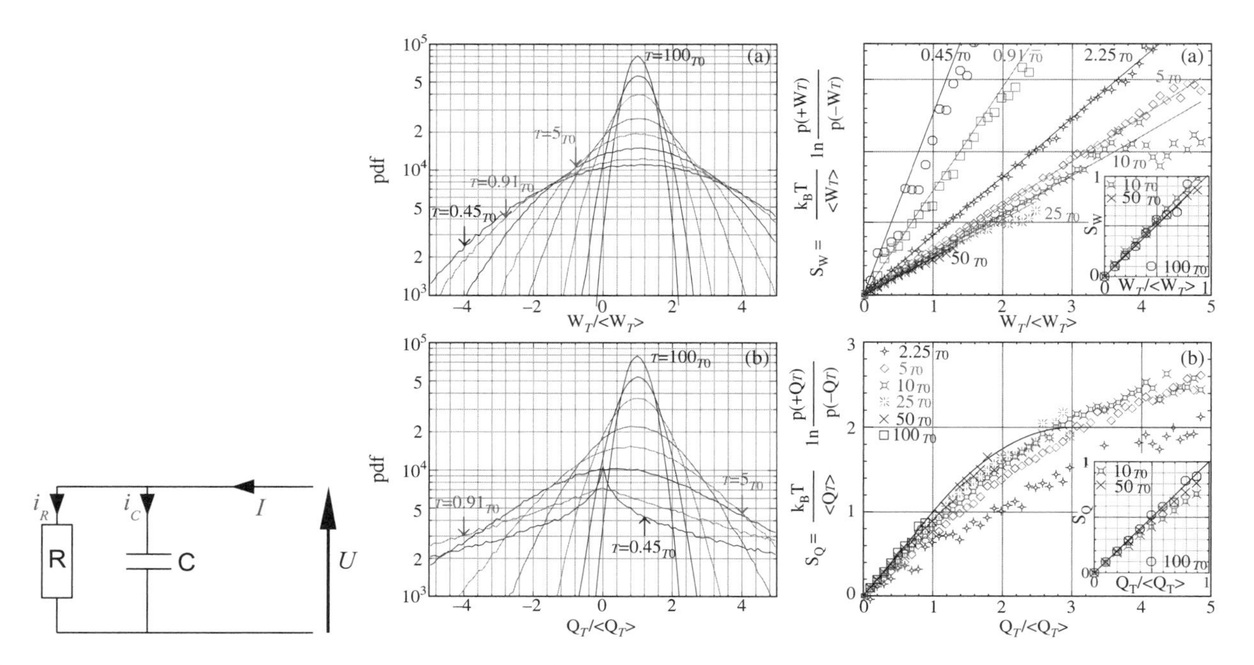

Figure 4. Heat and work fluctuations in an electrical circuit (left). PDF distributions (center) and verification of the FTs (Eqs. (83) and (84)) (right). (From Ref. 68.) (See color insert.)

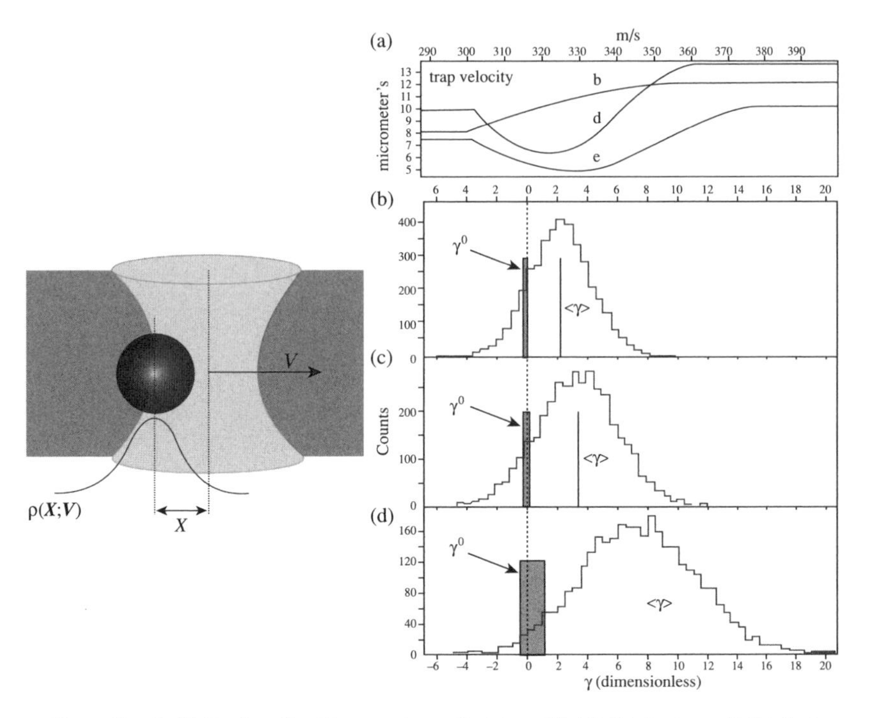

Figure 5. (Left) Bead confined in a moving optical trap. (Right) Total entropy $\mathcal{S}$ distributions (b–d) for the velocity protocols shown in (a). (From Ref. 69.) (See color insert.)

Let us now consider a NESS where the trap is moved at constant velocity, $x^*(t) = vt$. It is not possible to solve the Fokker–Planck equation to find the probability distribution in the steady state for arbitrary potentials. Only for harmonic potentials, $U(x) = \kappa x^2/2$, can the Fokker–Planck equation be solved exactly. The result is

$$P_{x^*}^{\mathrm{ss}}(x) = \left(\frac{2\pi T}{\kappa}\right)^{-1/2} \exp\left(-\frac{\kappa(x - x^*(t) + \gamma v/\kappa)^2}{2T}\right) \tag{60}$$

Note that the steady-state solution, Eq. (60), depends explicitly on time through $x^*(t)$. To obtain a time-independent solution we must change variables $x \to x - x^*(t)$ and describe the motion of the bead in the reference frame that is solid and moves with the trap. We will come back to this problem in Section IV.A.3.

1. Microscopic Reversibility

In this section we show that the Langevin dynamics, Eq. (57), satisfies the microscopic reversibility assumption or local detailed balance, Eq. (8). We recall that x is the position of the bead in the laboratory frame. The transition rates

$\mathcal{W}_{x^*}(x \to x')$ for the configuration x at time t to change to x' at a later time $t + \Delta t$ can be computed from Eq. (57). We discretize the Langevin equation [66] by writing

$$x' = x + \frac{f(x - x^*)}{\gamma}\Delta t + \sqrt{\frac{2T\,\Delta t}{\gamma}}r + \mathcal{O}((\Delta t)^2) \tag{61}$$

where r is a random Gaussian number of zero mean and unit variance. For a given value of x, the distribution of values x' is also a Gaussian with average and variance given by

$$\overline{x'} = x + \frac{f(x - x^*(t))}{\gamma}\Delta t + \mathcal{O}\Big((\Delta t)^2\Big) \tag{62}$$

$$\sigma_{x'}^2 = \overline{(x')^2} - (\overline{(x')})^2 = \frac{2T\,\Delta t}{\gamma} + \mathcal{O}\Big((\Delta t)^2\Big) \tag{63}$$

and therefore,

$$\mathcal{W}_{x^*}(x \to x') = (2\pi\sigma_{x'}^2)^{-1/2}\exp\left(-\frac{\left(x' - x + \frac{f(x - x^*)\Delta t}{\gamma}\right)^2}{2\sigma_{x'}^2}\right) \tag{64}$$

From Eq. (64) we compute the ratio between the transition probabilities to first order in Δt:

$$\frac{\mathcal{W}_{x^*}(x \to x')}{\mathcal{W}_{x^*}(x' \to x)} = \exp\left(-\frac{(x' - x)(f(x - x^*) + f(x' - x^*))}{2T}\right) \tag{65}$$

We can now use the Taylor expansions,

$$U(x' - x^*) = U(x - x^*) - f(x - x^*)(x' - x) + \mathcal{O}\Big((x' - x)^2\Big) \tag{66}$$

$$U(x - x^*) = U(x' - x^*) - f(x' - x^*)(x - x') + \mathcal{O}\Big((x' - x)^2\Big) \tag{67}$$

and subtract both equations to finally obtain

$$(x' - x)(f(x - x^*) + f(x' - x^*)) = 2(U(x' - x^*) - U(x - x^*)) \tag{68}$$

which yields

$$\frac{\mathcal{W}_{x^*}(x \to x')}{\mathcal{W}_{x^*}(x' \to x)} = \exp\left(-\frac{U(x' - x^*) - U(x - x^*)}{T}\right) = \frac{P_{x^*}^{\mathrm{eq}}(x')}{P_{x^*}^{\mathrm{eq}}(x)} \tag{69}$$

which is the local detailed balance assumption, Eq. (8).

2. *Entropy Production, Work, and Total Dissipation*

Let us consider an arbitrary nonequilibrium protocol $x^*(t)$, where $v(t) = \dot{x}^*(t)$ is the velocity of the moving trap. The entropy production for a given path, $\Gamma \equiv \{x(s); 0 \leq s \leq t\}$, can be computed using Eq. (22),

$$S_{\rm p}(\Gamma) = \int_0^t ds\, \dot{x}(s) \left(\frac{\partial \log P^{\rm eq}_{x^*(s)}(x)}{\partial x} \right)_{x=x(s)} \tag{70}$$

We now define the variable $y(t) = x(t) - x^*(t)$. From Eq. (59) we get[2]

$$S_{\rm p}(\Gamma) = \frac{1}{T}\int_0^t ds\, \dot{x}(s) f(x(s) - x^*(s)) = \frac{1}{T}\int_0^t ds(\dot{y}(s) + v(s)) f(y(s)) \tag{71}$$

$$= \frac{1}{T}\left(\int_{y(0)}^{y(t)} dy f(y) + \int_0^t ds\, v(s) f(s) \right) = \frac{-\Delta U + W(\Gamma)}{T} \tag{72}$$

with

$$\Delta U = U(x(t) - x^*(t)) - U(x(0) - x^*(0)); \quad W(\Gamma) = \int_0^t ds\, v(s) f(s) \tag{73}$$

where we used Eq. (58) in the last equality of Eq. (72). ΔU is the variation of internal energy between the initial and final positions of the bead and $W(\Gamma)$ is the mechanical work done by the moving trap on the bead. Using the first law, $\Delta U = W - Q$, we get

$$S_{\rm p}(\Gamma) = \frac{Q(\Gamma)}{T} \tag{74}$$

and the entropy production is just the heat transferred from the bead to the bath divided by the temperature of the bath.

The total dissipation $\mathcal{S}$, Eq. (23), can be evaluated by adding the boundary term, Eq. (24), to the entropy production. For the boundary term we have some freedom as to which function f we use on the rhs of Eq. (24):

$$B(\Gamma) = \log\big(P_{x^*(0)}(x(0))\big) - \log(f(x(t))) \tag{75}$$

[2]Note that $\dot{x}$, the velocity of the bead, is not well defined in Eqs. (70) and (72). However, $ds\,\dot{x}(s) = dx$ is. Yet, we prefer to use the notation in terms of velocities just to make clear the identification between the time integrals in Eqs. (70) and (72) and the discrete time-step sum in Eq. (22).

Because we want $\mathcal{S}$ to be antisymmetric against time reversal, there are two possible choices for the function f depending on the initial state.

- **Nonequilibrium Transient State (NETS).** Initially the bead is in equilibrium and the trap is at rest in a given position $x^*(0)$. Suddenly the trap is set in motion. In this case $b(x) = P^{\rm eq}_{x^*(t)}(x)$ and the boundary term in Eq. (24) reads

$$B(\Gamma) = \log\left(P^{\rm eq}_{x^*(0)}(x(0))\right) - \log\left(P^{\rm eq}_{x^*(t)}(x(t))\right) \tag{76}$$

By inserting Eq. (59) we obtain

$$B(\Gamma) = \frac{1}{T}(U(x(t) - x^*(t)) - U(x(0) - x^*(0))) = \frac{\Delta U}{T} \tag{77}$$

and $\mathcal{S} = S_{\rm p} + B = (Q + \Delta U)/T = W/T$ so the work satisfies the nonequilibrium equality, Eq. (16), and the FT, Eq. (27):

$$\frac{P_{\rm F}(W)}{P_{\rm R}(-W)} = \exp\left(\frac{W}{T}\right) \tag{78}$$

Note that in the reverse process the bead starts in equilibrium at the final position $x^*(t)$ and the motion of the trap is reversed $(x^*)^{\rm R}(s) = x^*(t-s)$. The result Eq. (78), is valid for arbitrary potentials $U(x)$. In general, the reverse work distribution $P_{\rm R}(W)$ will differ from the forward distribution $P_{\rm F}(W)$. Only for symmetric potentials $U(x) = U(-x)$ are both work distributions identical [67]. Under this additional assumption, Eq. (78) reads

$$\frac{P(W)}{P(-W)} = \exp\left(\frac{W}{T}\right) \tag{79}$$

Note that this is a particular case of the CFT (Eq. (41)) with $\Delta F = 0$.

- **Nonequilibrium Steady State (NESS).** If the initial state is a steady state, $P_{\lambda_0}(\mathcal{C}_0) \equiv P^{\rm ss}_{x^*(0)}(x)$, then we choose $b(x) = P^{\rm ss}_{x^*(t)}(x)$. The boundary term reads

$$B(\Gamma) = \log\left(P^{\rm ss}_{x^*(0)}(x(0))\right) - \log\left(P^{\rm ss}_{x^*(t)}(x(t))\right) \tag{80}$$

Only for harmonic potentials do we exactly know the steady-state solution, Eq. (60), so we can write down an explicit expression for B:

$$B(\Gamma) = \frac{\Delta U}{T} - \frac{v\gamma\,\Delta f}{\kappa T} \tag{81}$$

where ΔU is defined in Eq. (73) and $= f_{x^*(t)}(x(t)) - f_{x^*(0)}(x(0))$. The total dissipation is given by

$$\mathcal{S} = S_{\mathrm{p}} + B = \frac{Q + \Delta U}{T} - \frac{v\gamma\,\Delta f}{\kappa T} = \frac{W}{T} - \frac{v\gamma\,\Delta f}{\kappa T} \tag{82}$$

It is important to stress that Eq. (82) does not satisfy Eqs. (48) and (49) because the last boundary term on the rhs of Eq. (82) ($v\gamma\,\Delta f/\kappa T$) is not antisymmetric against time reversal. Van Zon and Cohen [61–63] have analyzed in much detail work and heat fluctuations in the NESS. They find that work fluctuations satisfy the exact relation

$$\frac{P(W)}{P(-W)} = \exp\left(\frac{W}{T}\frac{1}{1 + (\tau/t)(\exp(-t/\tau) - 1)}\right) \tag{83}$$

where t is the time window over which work is measured and τ is the relaxation time of the bead in the trap, $\tau = \gamma/\kappa$. Note that the FT (Eq. (79)) is satisfied in the limit $\tau/t \to 0$. Corrections to the FT are on the order of τ/t as expected (see discussion in the last part of Section III.C.2). Computations can also be carried out for heat fluctuations. The results are expressed in terms of the relative fluctuations of the heat, $a = S_{\mathrm{p}}/\langle S_{\mathrm{p}}\rangle$. The large deviation function $f_t(a)$ (Eq. (53)) is given by

$$\begin{aligned} &\lim_{t\to\infty} f_t(a) = a \quad (0 \le a \le 1) \\ &\lim_{t\to\infty} f_t(a) = a - (a-1)^2/4 \quad (1 \le a < 3) \\ &\lim_{t\to\infty} f_t(a) = 2 \quad (3 \le a) \end{aligned} \tag{84}$$

and $f_t(-a) = -f_t(a)$. Very accurate experiments to test Eqs. (83) and (84) have been carried out by Garnier and Ciliberto, who measured the Nyquist noise in an electric resistance [68]. Their results are in very good agreement with the theoretical predictions, which include corrections in the convergence of Eq. (84) on the order of $1/t$ as expected. A few results are shown in Fig. 4.

3. *Transitions Between Steady States*

Hatano and Sasa [56] have derived an interesting result for nonequilibrium transitions between steady states. Despite the generality of the Hatano–Sasa approach, explicit computations can be worked out only for harmonic traps. In the present example the system starts in a steady state described by the stationary distribution of Eq. (60) and is driven away from that steady state by varying the speed of the trap, v. The stationary distribution can be written in the frame system that moves solidly with the trap. If we define $y(t) = x - x^*(t)$ then Eq. (60) becomes

$$P_v^{\mathrm{ss}}(y) = \left(\frac{2\pi T}{\kappa}\right)^{-1/2} \exp\left(-\frac{\kappa(y + \gamma v/\kappa)^2}{2T}\right) \tag{85}$$

Note that, when expressed in terms of the reference moving frame, the distribution in the steady state becomes stationary or time independent. The transition rates in Eq. (64) can also be expressed in the reference system of the trap:

$$\mathcal{W}_v(y \to y') = \left(\frac{4\pi T \Delta t}{\gamma}\right)^{-1/2} \exp\left(-\frac{\gamma(y' - y + (v + \kappa/\gamma y)\Delta t)^2}{4T\,\Delta t}\right) \tag{86}$$

where we have used $f(x - x^*) = f(y) = -\kappa y$. The transition rates $\mathcal{W}_v(y \to y')$ now depend on the velocity of the trap. This shows that, for transitions between steady states, $\lambda \equiv v$ plays the role of the control parameter, rather than the value of x^*. A path is then defined by the evolution $\Gamma \equiv \{y(s); 0 \leq s \leq t\}$, whereas the perturbation protocol is specified by the time evolution of the speed of the trap $\{v(s); 0 \leq s \leq t\}$.

The rates $\mathcal{W}_v(y \to y')$ satisfy the local detailed balance property (Eq. (47)). From Eqs. (86) and (85), we get (in the limit $\Delta t \to 0$)

$$\frac{\mathcal{W}_v(y \to y')}{\mathcal{W}_v(y' \to y)} = \frac{P_v^{\mathrm{ss}}(y')}{P_v^{\mathrm{ss}}(y)} \tag{87}$$

$$= \exp\left(-\frac{\kappa}{2T}(y'^2 - y^2) - \frac{\gamma v}{T}(y' - y)\right) = \exp\left(-\left(\frac{\Delta U}{T} - \frac{v\gamma \Delta f}{\kappa T}\right)\right) \tag{88}$$

Note that the exponent on the rhs of Eq. (88) is equal to the boundary term, Eq. (81). In the reference system of the trap, we can then compute the entropy production S_{p} and the total dissipation $\mathcal{S}$. From either Eq. (22) or (88) and using Eq. (85), we get

$$S_{\mathrm{p}}(\Gamma) = \int_0^t ds\, \dot{y}(s) \left(\frac{\log(P_v^{\mathrm{ss}}(y))}{\partial y}\right)_{y=y(s)} = -\frac{\Delta U}{T} + \frac{\gamma}{\kappa T}\int_0^t ds\, v(s)\dot{F}(s) \tag{89}$$

$$= -\frac{1}{T}\left(\Delta U - \frac{\gamma}{\kappa}(\Delta(vF)) + \frac{\gamma}{\kappa}\int_0^t ds\, \dot{v}(s)F(s)\right) \tag{90}$$

where we integrated by parts in the last step of the derivation. For the boundary term, Eq. (56), we get

$$B(\Gamma) = \log\left(P_{v(0)}^{\mathrm{ss}}(y(0))\right) - \log\left(P_{v(t)}^{\mathrm{ss}}(y(t))\right) \tag{91}$$

$$= \frac{1}{T}\left(\Delta U - \frac{\gamma}{\kappa}(\Delta(vF)) + \frac{\gamma^2 \Delta(v^2)}{2\kappa}\right) \tag{92}$$

where we used Eq. (85). By adding Eqs. (90) and (92) we obtain the total dissipation,

$$\mathcal{S} = S_{\mathrm{p}} + B = \frac{1}{T}\left(\Delta\left(\frac{\gamma^2 v^2}{2\kappa}\right) - \frac{\gamma}{\kappa}\int_0^t ds\,\dot{v}(s)F(s)\right) \tag{93}$$

$$= -\frac{\gamma}{\kappa}\int_0^t ds\,\dot{v}(s)(F(s) - \gamma v(s)) \tag{94}$$

The quantity S (called Y by Hatano and Sasa) satisfies the nonequilibrium equality, Eq. (16), and the FT, Eq. (27). Only for time-reversal invariant protocols, $v^{\mathrm{R}}(s) = v(t-s)$, do we have $P_{\mathrm{F}}(\mathcal{S}) = P_{\mathrm{R}}(\mathcal{S})$, and the FT, Eq. (49), is also valid. We emphasize two aspects of Eq. (94).

- **Generalized Second Law for Steady-State Transitions.** From the inequality in Eqs. (18) and (94), we obtain

$$\frac{\gamma}{\kappa}\int_0^t ds\,\dot{v}(s)F(s) \leq \Delta\left(\frac{\gamma^2 v^2}{2\kappa}\right) \tag{95}$$

  which is reminiscent of the Clausius inequality $Q \geq -T\,\Delta S$, where the average dissipation rate $\overline{P}_{\mathrm{diss}} = \gamma v^2$ plays the role of a state function similar to the equilibrium entropy. In contrast to the Clausius inequality, the transition now occurs between steady states rather than equilibrium states [69].

- **Noninvariance of Entropy Production Under Galilean Transformations.** In steady states where $\dot{v} = 0$, S_{p} (Eq. (90)) becomes a boundary term and $\mathcal{S} = 0$ (Eq. (94)). We saw in Eq. (74) that S_{p} is equal to the heat delivered to the environment (and therefore proportional to the time elapsed t), whereas now S_{p} is a boundary term that does not grow with t. This important difference arises from the fact that the entropy production is not invariant under Galilean transformations. In the reference of the moving trap, the bath is moving at a given speed, which impedes one from defining heat in a proper way. To evaluate the entropy production for transitions between steady states, one has to resort to the description where x^* is the control parameter and $x^*(t) = \int_0^t ds\, v(s)$ is the perturbation protocol. In such a description, Eqs. (73) and (74) are still valid.

These results have been experimentally tested for trapped beads accelerated with different velocity protocols [69]. The results are shown in Fig. 5.

B. A Biological System: Pulling Biomolecules

The development of accurate instruments capable of measuring forces in the piconewton range and extensions on the order of the nanometer give access to a wide range of phenomena in molecular biology and biochemistry, where nonequilibrium processes that involve small energies on the order of a few k_BT are measurable (see Section II). From this perspective the study of biomolecules is an excellent playground to explore nonequilibrium fluctuations. The most successful investigations in this area have been achieved in single molecule experiments using optical tweezers [70]. In these experiments biomolecules can be manipulated one at a time by applying mechanical force at their ends. This allows us to measure small energies under varied conditions, opening new perspectives in the understanding of fundamental problems in biophysics (e.g., the folding of biomolecules) [71–73]. The field of single molecule research is steadily growing with new molecular systems being explored that show nonequilibrium behavior characteristic of small systems. The reader interested in a broader view of the area of single molecule research should have a look at Ref. 74.

1. *Single Molecule Force Experiments*

In single molecule force experiments, it is possible to apply force on individual molecules by grabbing the ends and pulling them apart [75–78]. Examples of different ways in which mechanical force is applied to single molecules are shown in Fig. 6. In what follows we will consider single molecule force experiments using optical tweezers, although everything we say extends to other

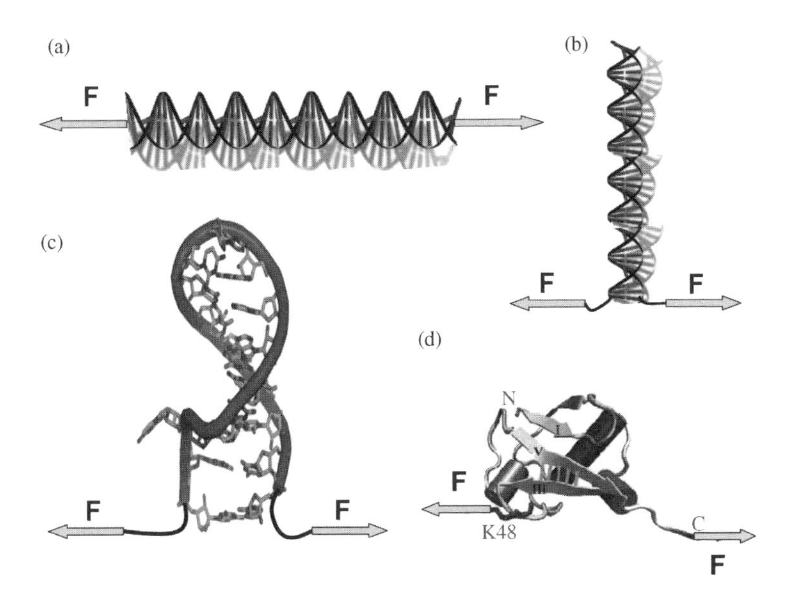

Figure 6. Pulling single molecules. (a) Stretching DNA; (b) unzipping DNA; (c) mechanical unfolding of RNA, and (d) mechanical unfolding of proteins. (See color insert.)

force techniques (AFM, magnetic tweezers, or biomembrane force probe; see Ref. 74) In these experiments, the ends of the molecule (e.g., DNA [79]) are labeled with chemical groups (e.g., biotin or digoxigenin) that can bind specifically to their complementary molecular partners (e.g., streptavidin or antidigoxigenin, respectively). Beads are then coated with the complementary molecules and mixed with the DNA in such a way that a tether connection can be made between the two beads through specific attachments. One bead is in the optical trap and used as a force sensor. The other bead is immobilized on the tip of a micropipette that can be moved by using a piezo-controlled stage to which the pipette is attached. The experiment consists of measuring force-extension curves (FECs) by moving the micropipette with respect to the trap position [80]. In this way it is possible to investigate the mechanical and elastic properties of the DNA molecule [81, 82].

Many experiments have been carried out by using this setup: the stretching of single DNA molecules, the unfolding of RNA molecules or proteins, and the translocation of molecular motors (Fig. 2). Here we focus our attention on force experiments where mechanical work can be exerted on the molecule and nonequilibrium fluctuations are measured. The most successful studies along this line of research are the stretching of small domain molecules such as RNA [83] or protein motifs [84]. Small RNA domains consist of a few tens of nucleotides folded into a secondary structure that is further stabilized by tertiary interactions. Because an RNA molecule is too small to be manipulated with micron-sized beads, it has to be inserted between molecular handles. These act as polymer spacers that avoid nonspecific interactions between the bead and the molecule as well as the contact between the two beads.

The basic experimental setting is shown in Fig. 7. We also show a typical FEC for an RNA hairpin and a protein. Initially, the FEC shows an elastic response due to the stretching of the molecular handles. Then, at a given value of the force, the molecule under study unfolds and a rip is observed in both force and extension. The rip corresponds to the unfolding of the small RNA/protein molecule. The molecule is then repeatedly stretched and relaxed, starting from the equilibrated native/extended state in the pulling/relaxing process. In the pulling experiment the molecule is driven out of equilibrium to a NETS by the action of a time-dependent force. The unfolding/refolding reaction is stochastic, the dissociation/formation of the molecular bonds that maintain the native structure of the molecule being determined by the Brownian motion of the surrounding water molecules [85]. Each time the molecule is pulled, different unfolding and refolding values of the force are observed (inset of Fig. 7b). The average value of the force at which the molecule unfolds during the pulling process increases with the loading rate (roughly proportional to the pulling speed) in a logarithmic way as expected for a two-state process (see discussion at the end of Section V.B.1 and Eq. (143)).

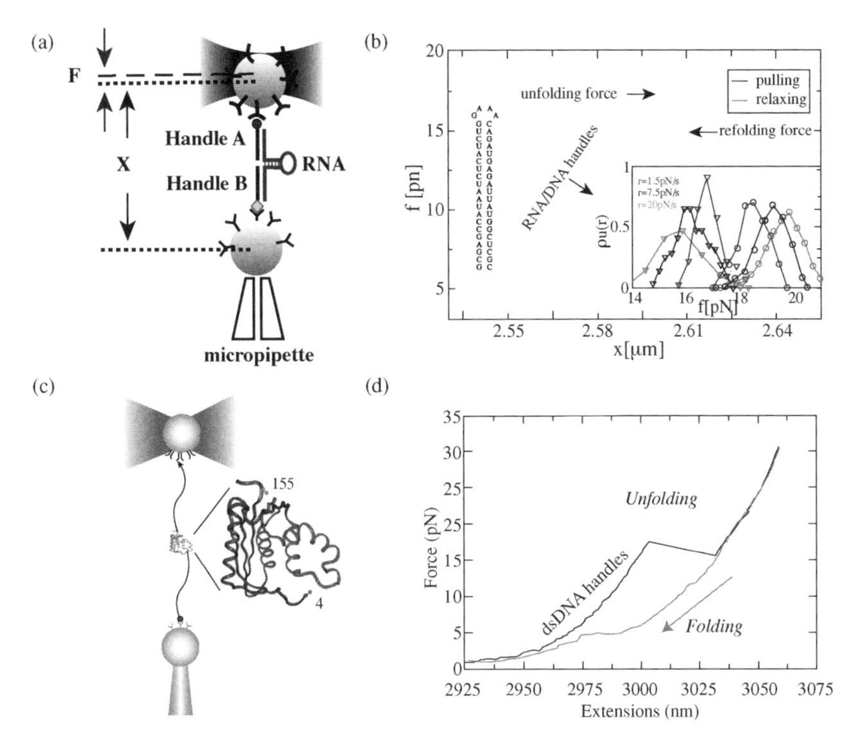

Figure 7. Mechanical unfolding of RNA molecules (a, b) and proteins (c, d) using optical tweezers. (a) Experimental setup in RNA pulling experiments. (b) Pulling cycles in the homologous hairpin and force rip distributions during the unfolding and refolding at three pulling speeds. (c) Equivalent setup in proteins. (d) Force extension curve when pulling the protein RNAseH. Panel (b) is from Ref. 86. Panels (a) and (d) are a courtesy from C. Cecconi [84]. (See color insert.)

Single molecule pulling experiments can be described with the formalism developed in Section III.C.1. In the simplest setting the configurational variable $\mathcal{C}$ corresponds to the molecular extension of the complex (handles plus inserted molecule) and the control parameter λ is either the force f measured in the bead or the molecular extension of the system, x. For small enough systems the thermodynamic equation of state is dependent on what is the variable that is externally controlled [87]. In the actual experiments, the assumption that either the force or the extension is controlled is just an approximation. Neither the molecular extension nor the force can be really controlled in optical tweezers [88]. For example, in order to control the force a feedback mechanism must operate at all times. This feedback mechanism has a time delay response so the force is never really constant [89, 90]. By assuming that the force is constant, we are neglecting some corrections in the analysis.[3] Under some conditions these corrections are shown to be unimportant (see below). Let us now consider that the force acting on the inserted molecule is controlled (the so-called isotensional ensemble). For a molecule that is repeatedly pulled from a minimum force value, $f_{\min}$, up to a maximum force, $f_{\max}$, the work (Eq. (37)) along a given path is given by

$$W_f(\Gamma) = \int_{f_{\min}}^{f_{\max}} df \frac{\partial E(x,f)}{\partial f} = -\int_{f_{\min}}^{f_{\max}} x(f)\, df \tag{96}$$

where the energy function is given by $E(x,f) = E(x,0) - fx$, $E(x,0)$ being the energy function of the molecule at zero force. The subindex f in W_f is written to underline the fact that we are considering the isotensional case where f is the control parameter. The Jarzynski equality, Eq. (40), and the FT, Eq. (41), hold with ΔF equal to the free energy difference between the initial and final equilibrium states. We assume that the molecule is immersed in water at constant temperature T and pressure p and acted on by a force f. The thermodynamic free energy $F(T,p,f)$ in this description is the Legendre transform of the Gibbs free energy at zero force, ambient temperature T, and pressure p, $G(T,p)$ [92, 93]:

$$F(T,p,f) = G(T,p) - fx(T,p,f) \rightarrow x(T,P,f) = -\frac{\partial F(T,p,f)}{\partial f} \tag{97}$$

We are interested in knowing the Gibbs free energy difference at zero force, ΔG, rather than the free energy difference ΔF between the folded state at $f_{\min}$ and the unfolded extended state at $f_{\max}$. We can express Eqs. (40) and (41) in terms of G (rather than F) and define the corrected work $W_f^c(\Gamma)$ along a path,

$$W_f^c(\Gamma) = W_f(\Gamma) + \Delta(xf) = W_f(\Gamma) + (x_{\max}f_{\max} - x_{\min}f_{\min}) \tag{98}$$

where the extensions $x_{\min}, x_{\max}$ are now fluctuating quantities evaluated at the initial and final times along each pulling. The corrected work $W_f^c(\Gamma)$ includes an additional boundary term and therefore does not satisfy either the JE or the CFT. If we now consider that x is the control parameter then we can define the equivalent of Eq. (96) (the so-called isometric ensemble):

$$W_x(\Gamma) = W_f(\Gamma) + \Delta(xf) = \int_{x_{\min}}^{x_{\max}} f(x)\, dx \tag{99}$$

[3] By using two traps, it is possible to maintain a constant force [91]. This is also possible with magnetic tweezers. However, because of the low stiffness of the magnetic trap, the spatial resolution due to thermal fluctuations is limited to a few tens of nanometers.

where now x is controlled and $x_{\min}, x_{\max}$ are fixed by the pulling protocol. Equations (98) and (99) look identical; however, they refer to different experimental protocols. Note that the term $W_f(\Gamma)$ appearing in Eq. (99) is now evaluated between the initial and final forces at fixed initial and final times. Both works W_x and W_f satisfy the relations (40) and (41). For a reversible process where f is controlled we have $W_f^{\text{rev}} = \Delta F$, whereas if x is controlled we have $W_x^{\text{rev}} = \Delta G$. In experiments it is customary to use Eq. (99) for the work: first, because that quantity is more easily recognized as the mechanical work; and second, because it gives the free energy difference between the folded and the unfolded states at zero force, a quantity that can be compared with thermal denaturation experiments.

In general, neither the force nor the molecular extension can be controlled in the experiments so definitions in Eqs. (96), (98), and (99) result in approximations to the *true* mechanical work that satisfy Eqs. (40) and (41). The control parameter in single molecule experiments using optical tweezers is the distance between the center of the trap and the immobilized bead [88]. Both the position of the bead in the trap and the extension of the handles are fluctuating quantities. It has been observed [94–96] that in pulling experiments the proper work that satisfies the FT includes some corrections to Eqs. (97) and (99) mainly due to the effect of the trapped bead. There are two considerations to take into account when analyzing experimental data.

- **W_x or W_f?** Let us suppose that f is the control parameter. In this case the JE and CFT, Eqs. (40) and (41), are valid for the work, Eq. (96). How large is the error that we make when we apply the JE using W_x instead? This question has been experimentally addressed by Ciliberto and co-workers [97, 98], who measured the work in an oscillator system with high precision (within tenths of $k_B T$). As shown in Eq. (99), the difference between both works is mainly a boundary term, $\Delta(xf)$. Fluctuations of this term can be a problem if they are on the same order as fluctuations of W_x itself. For a harmonic oscillator of stiffness constant equal to κ, the variance of fluctuations in fx are equal to $\kappa\delta(x^2)$, that is, approximately on the order of $k_B T$ due to the fluctuation-dissipation relation. Therefore, for experimental measurements that do not reach such precision, W_x or W_f is equally good.
- **The Effect of the Bead or Cantilever.** Hummer and Szabo [94] have analyzed the effect of a force sensor attached to the system (i.e., the bead in the optical trap or the cantilever in the AFM) in the work measurements. To this end, we consider a simplified model of the experimental setup (Fig. 8). In such a model, the molecular system (that includes the molecule of interest—RNA or protein— and the handles) is connected to a spring (that models the trapped bead or the AFM

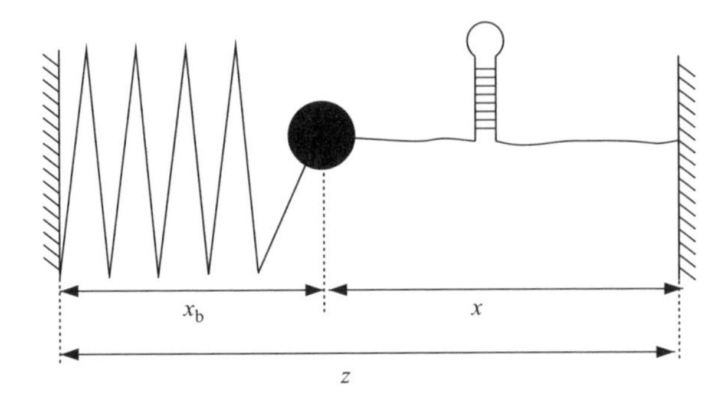

Figure 8. A molecular system of extension x is connected at its leftmost end to a bead trapped in an optical well (or to the tip of an AFM cantilever) and at its rightmost end to an immobilized surface (or a bead fixed to the tip of a micropipette). The position of the bead relative to the center of the trap, x_b, gives a readout of the acting force $f = \kappa x_b$. The control parameter in this setup is $z = x_b + x$, whereas both x_b and x are fluctuating quantities.

cantilever) and the whole system is embedded in a thermal bath. The total extension of the molecular system is x but the control parameter is $z = x + x_b$ where x_b is the position of the bead with respect to the center of the trap. The total free energy of the system is given by $F(x) + \frac{1}{2}\kappa x_b^2$, where $F(x)$ is the free energy of the molecular system alone and κ is the stiffness of the trap. The molecular extension x and the distance x_b are related by the force balance equation,

$$f = \kappa x_b = \frac{\partial F(x)}{\partial x} \tag{100}$$

where we assume that the bead is locally equilibrated at all values of the nonequilibrium molecular extension x (this is a good approximation as the bead relaxes fast enough compared to the typical time for the unfolding/refolding of the molecule). The mechanical work, Eq. (37), is then given by

$$W(\Gamma) = \int_{z_{\min}}^{z_{\max}} f\,dz = W_x(\Gamma) + \Delta\left(\frac{f^2}{2\kappa}\right) \tag{101}$$

where we used $dz = dx_b + dx$ and Eq. (100). The difference between the proper work W and W_x is again a boundary term. Because z is the control parameter, the JE and the CFT are valid for the work W but not for W_x. Again, the FT will not hold if fluctuations in the boundary term are important. The variance of these fluctuations is given by

$$\left\langle \delta\left(\frac{f^2}{2\kappa}\right)\right\rangle \approx \frac{k_B T\kappa}{\kappa_x + \kappa} \leq k_B T \tag{102}$$

where κ_x is the stiffness of the molecular system [88, 99]. Usually, $\kappa_x \gg \kappa$ so fluctuations in the boundary term are again smaller than $k_B T$. In general, as a rule of thumb, we can say that it does not matter much which mechanical work we measure if we do not seek free energy estimates with an accuracy less than $k_B T$. This is true unless the bead (cantilever) does not equilibrate within the time scales of the experiments. This may be the case when κ is too low and Eq. (100) is not applicable.

2. *Free Energy Recovery*

As we already emphasized, the JE (Eq. (40)) and the FT (Eq. (41)) can be used to predict free energy differences. In single molecule experiments it is usually difficult to pull molecules in a reversible way due to drift effects in the instrument. It is therefore convenient to devise nonequilibrium methods (such as the JE or the CFT) to extract equilibrium free energy differences from data obtained in irreversible processes. The first experimental test of the JE was carried out by pulling RNA hairpins that are derivatives of the L21 Tetrahymena ribozyme [100]. In these experiments RNA molecules were pulled at moderate speed: the average dissipated work in such experiments was less than $6k_B T$ and the work distributions turned out to be approximately Gaussian. Recent experiments have studied RNA molecules that are driven farther away from equilibrium in the nonlinear regime. In the nonlinear regime the average dissipated work is nonlinear with the pulling speed [101] and the work distribution strongly deviates from Gaussian [102]. In addition, these experiments have provided the first experimental test of the CFT (Eq. (41)). These measurements have also shown the possibility of recovering free energy differences by using the CFT with larger accuracy than that obtained by using the JE alone. There are two main predictions of the CFT (Eq. (41)) that have been scrutinized in these experiments.

- **Forward and Reverse Work Distributions Cross at $W = \Delta G$.** In order to obtain ΔG we can measure the forward and reverse work distributions, $P_F(W)$ and $P_R(-W)$, and look at the work value W^* where they cross, $P_F(W^*) = P_R(-W^*)$. According to Eq. (41), both distributions should cross at $W^* = \Delta G$ independently of how far the system is driven out of equilibrium (i.e., independently of the pulling speed). Figure 9 shows experiments on a short canonical RNA hairpin CD4 (i.e., just containing Watson–Crick complementary base pairs) at three different pulling speeds, which agree very well with the FT prediction.
- **Verification of the CFT.** The CFT (Eq. (41)) can be tested by plotting $\log(P_F(W)/P_R(-W))$ as a function of W. The resulting points should fall in a straight line of slope 1 (in $k_B T$ units) that intersects the work axis at

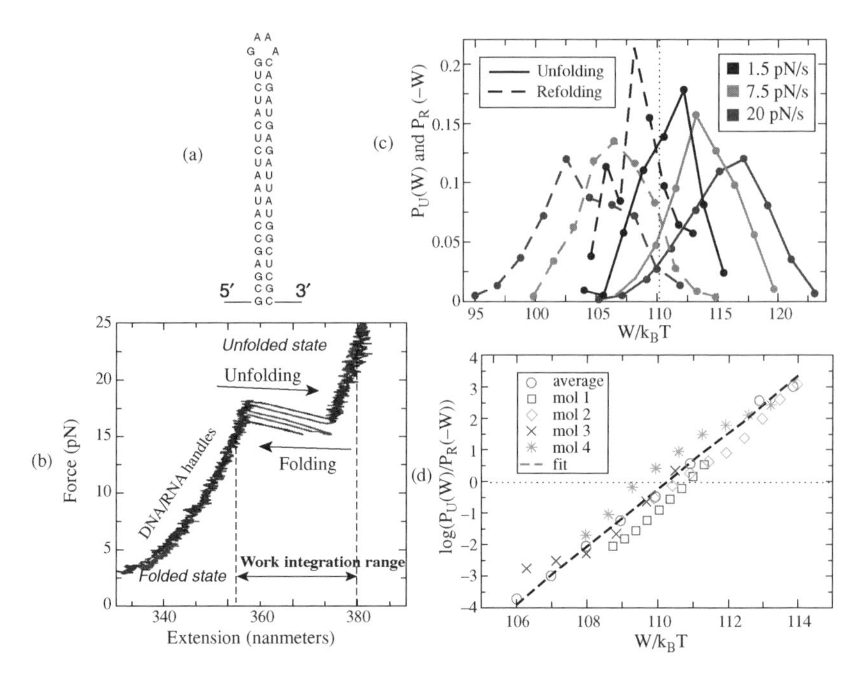

Figure 9. (a) Structure of the homologous CD4 hairpin. (b) FECs at a loading rate of 1.7 pN/s. (c) Unfolding and refolding work distributions at three loading rates (see inset). The unfolding and refolding work distributions cross at a value ΔG independent of the pulling speed as predicted by the CFT. Data correspond to 100,400 and 700 pulls for the lowest, medium, and highest pulling speeds, respectively. (d) Test of the CFT at the intermediate loading rate 7.5 pN/s for four different tethers. The trend of the data is reproducible from tether to tether and consistent with the CFT prediction. (From Ref. 102.) (See color insert.)

$W = \Delta G$. Of course, this relation can be tested only in the region of work values along the work axis where both distributions (forward and reverse) overlap. An overlap between the forward and reverse distributions is hardly observed if the molecules are pulled too fast or if the number of pulls is too small. In such cases, other statistical methods (Bennet's acceptance ratio or maximum likelihood methods, Section IV.B.3) can be applied to get reliable estimates of ΔG. The validity of the CFT has been tested in the case of the RNA hairpin CD4 previously mentioned and the three-way junction RNA molecule as well. Figure 9c,d and Fig. 10c show results for these two molecules.

In general, both the JE and the CFT are only valid in the limit of an infinite number of pulls. For a finite number of pulls, N, the estimated value for ΔG that

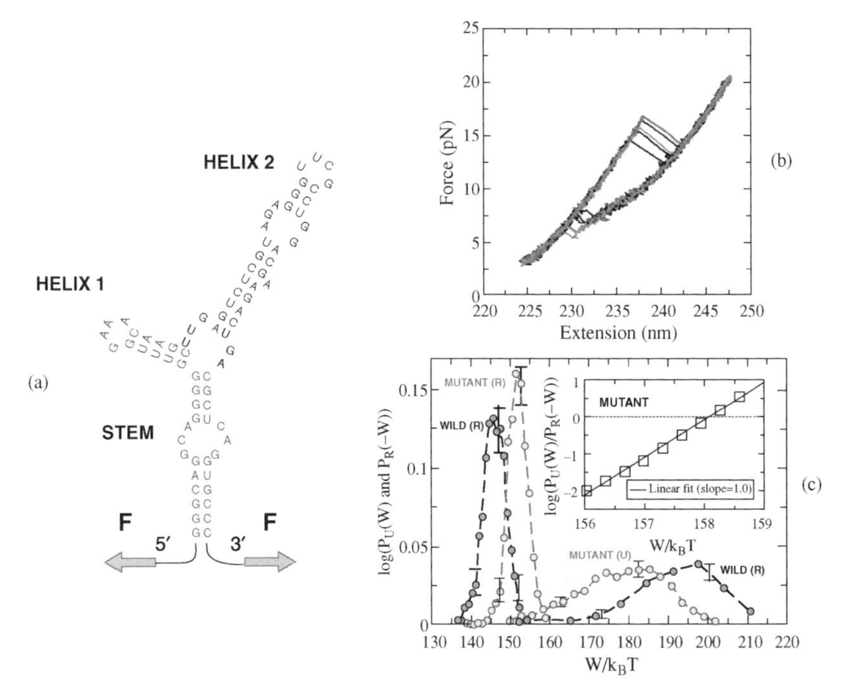

Figure 10. (a) Secondary structure of the three-way junction S15. (b) A few FECs for the wild type. (c) Unfolding/refolding work distributions for the wild type and the mutant. (Inset) Experimental verification of the validity of the CFT for the mutant, where unfolding and refolding distributions overlap each other over a limited range of work values. Data correspond to 900 pulls for the wild type and 1200 pulls for the mutant. (From Ref. 102.) (See color insert.)

is obtained by applying the JE is biased [103]. The free energy estimate F_k^{JE} for a given set, k, of N work values $W_1^k, W_2^k, \ldots, W_N^k$ is defined as

$$F_k^{\mathrm{JE}} = -T\log\left(\frac{1}{N}\sum_{i=1}^{N}\exp\left(-\frac{W_i^k}{T}\right)\right) \tag{103}$$

The free energy bias is defined by averaging the estimator F_k^{JE} over an infinite number of sets,

$$B(N) = \lim_{M\to\infty}\left(\frac{1}{M}\sum_{k=1}^{M} F_k^{\mathrm{JE}}\right) - \Delta F \tag{104}$$

where ΔF is the true free energy difference. The bias $B(N)$ converges to 0 for $N \to \infty$. However, it is of practical importance to devise methods to estimate how many pulls are required to obtain the Jarzynski free energy estimate F^{JE} within a reasonable error far from the true value [101, 104]. The bias is a complicated mathematical object because the Jarzynski average catches important contributions from large deviations of the work. As we will see in Section V.C.2, the bias is a large deviation function that requires specific mathematical methods to analyze its finite N behavior and large N convergence. There we prove that, for large N, the bias decreases as $1/N$, a result known as the Woods formula [104]. In the intermediate N regime, the behavior of the bias is more complicated [105]. Free energy recovery techniques are also used in numerical simulations to evaluate free energy differences [106–109] and reconstruct free energy profiles or potentials of mean-field force [110, 111].

3. *Efficient Strategies and Numerical Methods*

An important question is to understand the optimum nonequilibrium protocol to recover free energies using the JE given specific constraints in experiments and simulations. There are several considerations to take into account.

- **Faster or Slower Pulls?** In single molecule experiments, tethers break often so it is not possible to repeatedly pull the same tether an arbitrary number of times. Analogously, in numerical simulations only a finite amount of computer time is available and only a limited number of paths can be simulated. Given these limitations, is it better to perform many fast pulls or a few slower pulls to recover the free energy difference using the JE? In experiments, drift effects in the instrument always put severe limitations on the minimum speed at which molecules can be pulled. To obtain good quality data, it is advisable to carry out pulls as fast as possible. In numerical simulations, the question about the best strategy for free energy recovery has been considered in several papers [103, 112–114]. The general conclusion that emerges from these studies is that, in systems that are driven far away from equilibrium, it is preferable to carry out many pulls at high speed than a few pulls at slower speeds. The reason can be intuitively understood. Convergence in the JE is dominated by the so-called outliers, that is, work values that deviate a lot from the average work and are smaller than ΔF. The outliers contribute a lot to the exponential average, Eq. (40). For higher pulling speeds, we can perform more pulls so there are more chances to catch a large deviation event, that is to catch an outlier. At the same time, because at higher speeds the pulling is more irreversible, the average dissipated work becomes larger, making the free energy estimate less reliable. However, the contribution of the outliers required to recover the correct free energy is more important than the opposite effect due to the increase of the average dissipated work. We should mention that periodically oscillating pulls have also been considered; however, it is unclear whether they lead to improved free energy estimates [115, 116].

- **Forward or Reverse Process?** Suppose we want to evaluate the free energy difference between two states, A and B, by using the JE. Is it better to estimate ΔF by carrying out irreversible experiments from A to B, or is it better to do them from B to A? Intuitively, it seems natural that the less irreversible process among the two (forward and reverse), which is the one with smaller dissipated work W_{diss}, is also the most convenient to consider in order to extract the free energy difference. However, this is not true. In general, a larger average dissipated work implies a larger work variance (Eq. (42))—that is, larger fluctuations. The larger the fluctuations, the larger is the probability to catch a large deviation that contributes to the exponential average. It seems reasonable that if outliers contribute much more to finding the right free energy than proper tuning of the average value of the work, then the process that fluctuates more (i.e., the more dissipative one) is the process that must be sampled to efficiently recover ΔF. This result was anticipated in Ref. 117 and analyzed in more detail in Ref. 118. For Gaussian work distributions, the minimum number of pulls, N^*, required to efficiently recover free energy differences within $1k_{\mathrm{B}}T$ by using the JE grows exponentially with the dissipated work along the nonequilibrium process [101]. However, for general work distributions, the value of $N^*_{\mathrm{F(R)}}$ along the forward (reversed) process depends on the average dissipated work along the reverse (forward) process [118]. This implies that

$$N^*_{\mathrm{F(R)}} \sim \exp\left(\frac{W^{\mathrm{R(F)}}_{\mathrm{diss}}}{T}\right) \tag{105}$$

  and the process that dissipates most between the forward and the reverse is the best to efficiently recover ΔF.

Until now we discussed strategies for recovering free energy differences using the JE. We might be interested in free energy recovery by combining the forward and reverse distributions at the same time that we use the CFT. This is important in both experiments [102] and simulations [119, 120] where it is convenient and natural to use data from the forward and reverse processes. The best strategy to efficiently recover free energies using the forward and reverse processes was proposed by C. Bennett in the context of equilibrium sampling [121]. The method was later extended by Crooks to the nonequilibrium case [46] and is known as Bennett's acceptance ratio method. The basis of the method is as follows. Let us multiply both sides of Eq. (41) by the function $g_\mu(W)$,

$$g_\mu(W)\exp\left(-\frac{W}{T}\right)P_{\rm F}(W) = g_\mu(W)P_{\rm R}(-W)\exp\left(-\frac{\Delta F}{T}\right) \tag{106}$$

where $g_\mu(W)$ is an arbitrary real function that depends on the parameter μ. Integrating both sides between $W = -\infty$ and $W = \infty$ gives

$$\left\langle g_\mu(W)\exp\left(-\frac{W}{T}\right)\right\rangle_{\rm F} = \langle g_\mu(W)\rangle_{\rm R}\exp\left(-\frac{\Delta F}{T}\right) \tag{107}$$

where $\langle\cdots\rangle_{(\rm F,R)}$ denote averages over the forward and reverse process, respectively. Taking the logarithm of both sides, we have

$$z_{\rm R}(\mu) - z_{\rm F}(\mu) = \frac{\Delta F}{T} \tag{108}$$

where we have defined

$$z_{\rm R}(\mu) = \log(\langle g_\mu(W)\rangle_{\rm R}) \tag{109}$$

$$z_{\rm F}(\mu) = \left\langle g_\mu(W)\exp\left(-\frac{W}{T}\right)\right\rangle_{\rm F} \tag{110}$$

Equation (108) implies that the difference between functions $z_{\rm F}$ and $z_{\rm R}$ must be a constant over all μ values. The question we would like to answer is the following. Given a finite number of forward and reverse pulls, what is the optimum choice for $g_\mu(W)$ that gives the best estimate of Eq. (108) for ΔF? For a finite number of experiments $N_{\rm F}, N_{\rm R}$ along the forward and reverse process we can write

$$\langle A(W)\rangle_{\rm F(R)} = \frac{1}{N_{\rm F(R)}}\sum_{i=1}^{N_{\rm F(R)}} A(W_i) \tag{111}$$

for any observable A. Equation (107) yields an estimate for ΔF,

$$(\Delta F)^{\rm est} = T\left(\log(\langle g_\mu(W)\rangle_{\rm R}) - \log\left(\left\langle g_\mu(W)\exp\left(-\frac{W}{T}\right)\right\rangle_{\rm F}\right)\right) \tag{112}$$

Minimization of the variance,

$$\sigma^2_{\Delta F} = \left\langle\left((\Delta F)^{\rm est} - \Delta F\right)^2\right\rangle \tag{113}$$

($\langle \cdots \rangle$ denotes the average over the distributions $P_{\mathrm{F}}, P_{\mathrm{R}}$) with respect to all possible functions $g_{\mu}(W)$ shows [46, 121] that the optimal solution is given by

$$g_{\mu}(W) = \frac{1}{1 + (N_{\mathrm{F}}/N_{\mathrm{R}}) \exp((W - \mu)/T)} \tag{114}$$

and $\mu = \Delta F$. The same result has been obtained by Pande and co-workers by using maximum likelihood methods [122]. In this case, one starts from a whole set of work data encompassing N_{F} forward and N_{R} reversed values. One then defines the likelihood function of distributing all work values between the forward and reverse sets. Maximization of the likelihood leads to Bennett's acceptance ratio formula. To extract ΔF it is then customary to plot the difference on the lhs of Eq. (108), $z_{\mathrm{R}}(\mu) - z_{\mathrm{F}}(\mu)$, as a function of μ by using Eqs. (109) and (110). The intersection with the line $z_{\mathrm{R}}(\mu) - z_{\mathrm{F}}(\mu) = \mu$ gives the best estimate for ΔF. An example of this method is shown in Fig. 11. Recently, the maximum likelihood method has been generalized to predict free energy estimates between more than two states [123].

V. PATH THERMODYNAMICS

A. The General Approach

The JE (Eq. (40)) indicates a way to recover free energy differences by measuring the work along all possible paths that start from an equilibrium state. Its mathematical form reminds one of the partition function in the canonical ensemble used to compute free energies in statistical mechanics. The formulas for the two cases are

$$\sum_{C} \exp\left(-\frac{E(C)}{T}\right) = \exp\left(-\frac{F}{T}\right) \quad \text{(partition function)} \tag{115}$$

$$\sum_{\Gamma} \exp\left(-\frac{W(\Gamma)}{T}\right) = \exp\left(-\frac{\Delta F}{T}\right) \quad \text{(Jarzynski equality)} \tag{116}$$

where F is the equilibrium free energy of the system at temperature T. Throughout this section we take $k_{\mathrm{B}} = 1$. In the canonical ensemble the entropy $S(E)$ is equal to the logarithm of the density of states with a given energy E. That density is proportional to the number of configurations with energy equal to E. Therefore, Eq. (115) becomes

$$\begin{aligned} \exp\left(-\frac{F}{T}\right) &= \sum_{C} \exp\left(-\frac{E(C)}{T}\right) \\ &= \sum_{E} \exp\left(S(E) - \frac{E}{T}\right) = \sum_{E} \exp\left(-\frac{\Phi(E)}{T}\right) \end{aligned} \tag{117}$$

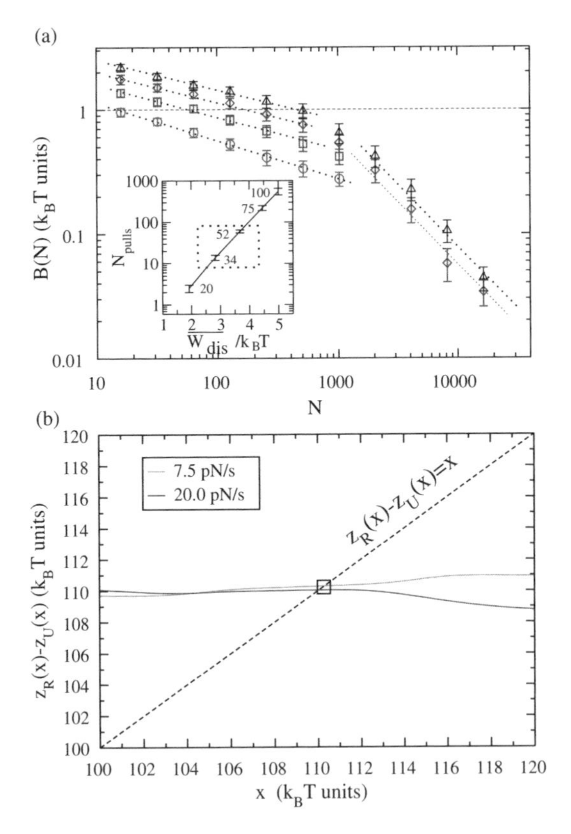

Figure 11. (a) Bias as a function of the number of pulls N for a two-states model. The inset shows the number of pulls as a function of the dissipated work required to recover the free energy with an error within $1k_BT$. (b) Function $z_R - z_F$ for the data shown in Fig. 9c at the two largest pulling speeds. Panel (a) from (Refs. 3 and 101; panel (b) from the supplementary material in Ref. 102.) (See color insert.)

where $\Phi(E) = E - TS$ is the free energy functional. In the large volume limit, the sum in Eq. (117) is dominated by the value $E = E^{\text{eq}}$, where $F(E)$ is minimum. The value E^{eq} corresponds to the equilibrium energy of the system and $\Phi(E^{\text{eq}})$ is the equilibrium free energy. The following relations hold:

$$F = \Phi(E^{\text{eq}}); \quad \left(\frac{\partial \Phi(E)}{\partial E}\right)_{E=E^{\text{eq}}} = 0 \rightarrow \left(\frac{\partial S(E)}{\partial E}\right)_{E=E^{\text{eq}}} = \frac{1}{T} \tag{118}$$

The equilibrium energy E^{eq} is different from the most probable energy, E^{mp}, defined by $S'(E = E^{\text{mp}}) = 0$. E^{mp} is the average energy we would find if we were to randomly select configurations all with identical *a priori* probability. The equilibrium energy, rather than the most probable energy, is the thermodynamic energy for a system in thermal equilibrium.

Proceeding in a similar way for the JE, we can define the *path entropy S(W)* as the logarithm of the density of paths with work equal to W, $P(W)$:

$$P(W) = \exp(S(W)) \tag{119}$$

We can rewrite Eq. (116) in the following way:

$$\begin{aligned}\exp\left(-\frac{\Delta F}{T}\right) &= \sum_\Gamma \exp\left(-\frac{W(\Gamma)}{T}\right) = \int dW\, P(W) \exp\left(-\frac{W}{T}\right) \\ &= \int dW \,\exp\left(S(W) - \frac{W}{T}\right) = \int dW \,\exp\left(-\frac{\Phi(W)}{T}\right)\end{aligned} \tag{120}$$

where $\Phi(W) = W - TS(W)$ is the *path free energy*. In the large volume limit, the sum in Eq. (120) is dominated by the work value, $W^\dagger$, where $\Phi(W)$ is minimum. Note that the value $W^\dagger$ plays the role of the equilibrium energy in the canonical case, Eq. (118). From Eq. (120) the path free energy $\Phi(W^\dagger)$ is equal to the free energy difference ΔF. The following relations hold:

$$\Delta F = \Phi(W^\dagger) = W^\dagger - TS(W^\dagger) \tag{121}$$

$$\left(\frac{\partial \Phi(W)}{\partial W}\right)_{W=W^\dagger} = 0 \rightarrow \left(\frac{\partial S(W)}{\partial W}\right)_{W=W^\dagger} = \frac{1}{T} \tag{122}$$

At the same time, $W^\dagger$ is different from the most probable work, W^{mp}, defined as the work value at which $S(W)$ is maximum:

$$\left(\frac{\partial S(W)}{\partial W}\right)_{W=W^{\mathrm{mp}}} = 0 \rightarrow \left(\frac{\partial \Phi(W)}{\partial W}\right)_{W=W^{\mathrm{mp}}} = 1 \tag{123}$$

The role of W^{mp} and $W^\dagger$ in the case of the JE (Eq. (115)) and E^{mp} and E^{eq} in the partition function case (Eq. (116)) appear exchanged. W^{mp} is the work value typically observed upon repetition of the same experiment a large number of times. In contrast, in the partition function case (Eq. (115)), E^{mp} is not the typical energy, the typical energy being E^{eq}. In addition, $W^\dagger$ is not the typical work but the work that must be sampled along paths in order to be able to extract the free energy difference using the JE. As we have already emphasized, as the system size increases, less and less paths can sample the region of work values around $W^\dagger$. Therefore, although both formalisms (partition function and JE) are mathematically similar, the physical meaning of the quantities $W^\dagger$ and E^{eq} is different. In the large volume limit, E^{eq} is almost *always* observed whereas $W^{\dagger}$ is almost *never* observed.

In general, from the path entropy we can also define a *path temperature*, $\hat{T}(W)$,

$$\frac{\partial S(W)}{\partial W} = \lambda(W) = \frac{1}{\hat{T}(W)} \rightarrow \hat{T}(W^{\dagger}) = T \tag{124}$$

where $\lambda(W)$ is a Lagrange multiplier that transforms the path entropy $S(W)$ into the path free energy $\Phi(W)$, Eq. (121). The mathematical relations between the new quantities $W^{\dagger}$ and W^{mp} can be graphically represented for a given path entropy $S(W)$. This is shown in Fig. 12.

The path thermodynamics formalism allows us to extract some general conclusions on the relation between $W^{\dagger}$ and W^{mp}. Let us consider the CFT

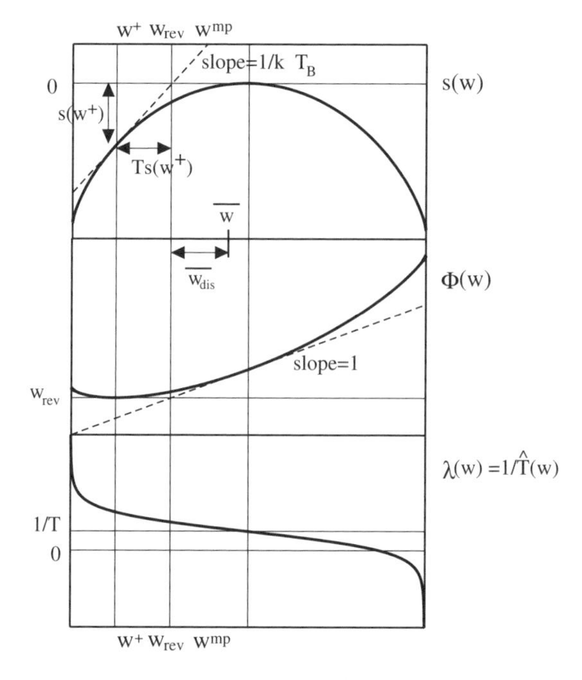

Figure 12. (Upper panel) Path entropy $s(w)$; (Middle panel) path free-energy $\Phi(w) = w - Ts(w)$; and (lower panel) Lagrange multiplier $\lambda(w)$ equal to the inverse of the path temperature $1/\hat{T}(w)$. w^{mp} is the most probable work value given by $s'(w^{\mathrm{mp}}) = \lambda(w^{\mathrm{mp}}) = 0$ or $\Phi'(w^{\mathrm{mp}}) = 1$; $w^{\dagger}$ is the value of the work that has to be sampled to recover free energies from nonequilibrium work values using the JE. This is given by $s'(w^{\dagger}) = 1/T$ or $\Phi'(w^{\dagger}) = 0$; w_{rev} and $\overline{w_{\mathrm{dis}}}$ are the reversible and average dissipated work, respectively. (From Ref. 117.)

(Eq. (41)). In terms of the path entropies for the forward and reverse processes, $S_F(W)$ and $S_R(W)$, (Eq. (41)) can be written as

$$S_F(W) - S_R(-W) = \frac{W - \Delta F}{T} \rightarrow (S_F)'(W) + (S_R)'(-W) = \frac{1}{T} \tag{125}$$

where we used Eq. (119) and later derived it with respect to W. By inserting $W = W_F^\dagger$ and $-W_R{}^\dagger$ on the rhs of Eq. (125) and using Eqs. (122) and (123), we obtain the following chain of relations:

$$(S_F)'(W_F^\dagger) + (S_R)'(-W_F^\dagger) = \frac{1}{T} \rightarrow (S_R)'(-W_F^\dagger) = 0 \rightarrow W_F^\dagger = -W_R^{\mathrm{mp}} \tag{126}$$

$$(S_F)'(-W_R^\dagger) + (S_R)'(W_R^\dagger) = \frac{1}{T} \rightarrow (S_F)'(-W_R^\dagger) = 0 \rightarrow W_R^\dagger = -W_F^{\mathrm{mp}} \tag{127}$$

The rightmost equalities in Eqs. (126) and (127) imply that the most probable work along the forward (reverse) process is equal to the work value ($W^\dagger$) that must be sampled, in a finite number of experiments, along the reverse (forward) process for the JE to be satisfied. This result has already been discussed in Section IV.B.3: the process that dissipates most between the forward and the reverse is the one that samples more efficiently the region of values close to $W^\dagger$. This conclusion, which may appear counterintuitive, can be rationalized by noting that larger dissipation implies larger fluctuations and therefore more chances to get rare paths that sample the vicinity of $W^\dagger$. The symmetries in Eqs. (126) and (127) were originally discussed in Ref. 117 and analyzed in detail for the case of the gas contained in a piston [118].

We close this section by analyzing the case where the work distribution is Gaussian. The Gaussian case describes the linear response regime usually (but not necessarily) characterized by small deviations from equilibrium. Let us consider the following distribution:

$$P(W) = (2\pi\sigma_W^2)^{-1/2} \exp\left(-\frac{(W - W^{\mathrm{mp}})^2}{2\sigma_W^2}\right) \tag{128}$$

where the average value of the work, $\langle W \rangle$, is just equal to the most probable value W^{mp}. The path entropy is given by $S(W) = -(W - W^{\mathrm{mp}})^2/(2\sigma_W^2) + \text{constant}$, so Eq. (123) is satisfied. From Eq. (124) we get

$$\hat{T}(W) = -\frac{\sigma_W^2}{W - W^{\mathrm{mp}}} \rightarrow W^\dagger = W^{\mathrm{mp}} - \frac{\sigma_W^2}{T} \tag{129}$$

From Eqs. (122) and (129) we get $W^\dagger = \Delta F - (\sigma_W^2/2T)$. Therefore,

$$W_{\rm diss}^\dagger = W^\dagger - W_{\rm rev} = W^\dagger - \Delta F = -\frac{\sigma_W^2}{2T} \tag{130}$$

$$W_{\rm diss}^{\rm mp} = W^{\rm mp} - W_{\rm rev} = W^{\rm mp} - \Delta F = \frac{\sigma_W^2}{2T} \tag{131}$$

leading to the final result $W_{\rm diss}^\dagger = -W_{\rm diss}^{\rm mp} = -\langle W_{\rm diss}\rangle$. Therefore, in order to recover the free energy using the JE, paths with negative dissipated work and of magnitude equal to the average dissipated work must be sampled. Sometimes the paths with negative dissipated work are referred to as *transient violations of the second law*. This name has raised strong objections among some physicists. Of course, the second law remains inviolate. The name just stresses the fact that paths with negative dissipated work must be experimentally accessible to efficiently recover free energy differences. Note that, for the specific Gaussian case, we get $\langle W_{\rm diss}\rangle = \sigma_W^2/2T$ therefore the fluctuation-dissipation parameter R (Eq. (42)) is equal to 1 as expected for systems close to equilibrium. The result $R = 1$ has been shown to be equivalent to the validity of the fluctuation-dissipation theorem [96].

B. Computation of the Work/Heat Distribution

The JE and the CFT describe relations between work distributions measured in the NETS. However, they do not imply a specific form of the work distribution. In small systems, fluctuations of the work relative to the average work are large so work distributions can strongly deviate from Gaussian distributions and be highly nontrivial. In contrast, as the system size increases, deviations of the work respect to the average value start to become rare and exponentially suppressed with the system size. To better characterize the pattern of nonequilibrium fluctuations, it seems important to explore analytical methods that allow us to compute, at least approximately, the shape of the energy distributions (e.g., heat or work) along nonequilibrium processes. Of course, there is always the possibility of carrying out exact calculations in specific solvable cases. In general, however, the exact computation of the work distribution can be a difficult mathematical problem (solvable examples are given in Refs. 124–128) that is related to the evaluation of large deviation functions (Section V.C). This problem has traditionally received a lot of attention by mathematicians and we foresee it may become a central area of research in statistical physics in the next few years.

1. An Instructive Example

To put the problem in proper perspective, let us consider an instructive example: an individual magnetic dipole of moment μ subject to a magnetic field H and embedded in a thermal bath. The dipole can switch between the up and down configurations, $\pm\mu$. The transition rates between the up and down orientations are of the Kramers type [129, 130],

$$k_{-\mu\to\mu}(H) = k_{\text{up}}(H) = k_0 \frac{\exp(\mu H/T)}{2\cosh(\mu H/T)} \tag{132}$$

$$k_{\mu\to-\mu}(H) = k_{\text{down}}(H) = k_0 \frac{\exp(-\mu H/T)}{2\cosh(\mu H/T)} \tag{133}$$

with $k_0 = k_{\text{up}}(H) + k_{\text{down}}(H)$ independent of H. The rates in Eqs. (132) and (133) satisfy detailed balance (Eq. (8)):

$$\frac{k_{\text{up}}(H)}{k_{\text{down}}(H)} = \frac{P^{\text{eq}}(\mu)}{P^{\text{eq}}(-\mu)} = \exp\left(\frac{2\mu H}{T}\right) \tag{134}$$

with $P^{\text{eq}}(-(+)\mu) = \exp(-(+)\mu H/T)/\mathcal{Z}$, where $\mathcal{Z} = 2\cosh(\mu H/T)$ is the equilibrium partition function. In this system there are just two possible configurations: $\mathcal{C} = -\mu, \mu$. We consider a nonequilibrium protocol where the control parameter H is varied as a function of time, $H(t)$. The dynamics of the dipole is a continuous time Markov process, and a path is specified by the time sequence $\Gamma \equiv \{\mu(t)\}$.

Let us consider the following protocol: the dipole starts in the down state $-\mu$ at $H = -H_0$. The field is then ramped from $-H_0$ to $+H_0$ at a constant speed $r = \dot{H}$, so $H(t) = rt$ (Fig. 13a). The protocol lasts for a time $t_{\text{max}} = 2H_0/r$ and the field stops changing when it has reached the value H_0. The free energy difference between the initial and final states is 0 because the free energy is an even function of H. To ensure that the dipole initially points down and that this is an equilibrium state, we take the limit $H_0 \to \infty$ but we keep the ramping speed r finite. In this way we generate paths that start at $H = -\infty$ at $t = -\infty$ and end up at $H = \infty$ at $t = \infty$. We can now envision all possible paths followed by the dipole. The up configuration is statistically preferred for $H > 0$,

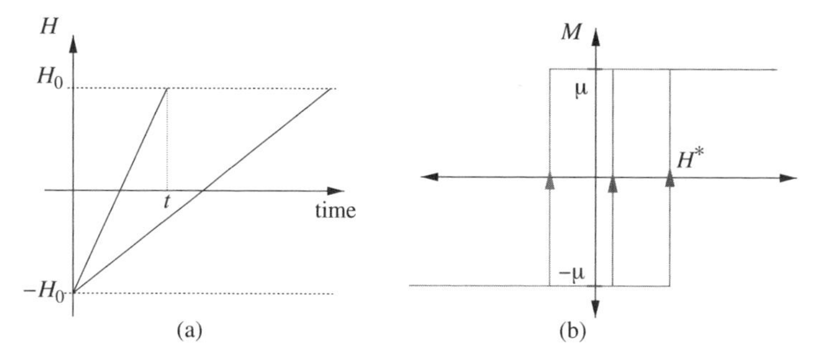

Figure 13. (a) Ramping protocol. The ramping speed is defined by $r = 2H_0/t$ where t is the duration of the ramp. (b) Three examples of paths where the down dipole reverses orientation at different values of the field, H^*. (See color insert.)

whereas the down configuration is preferred for $H < 0$. Therefore, in a typical path the dipole will stay in the down state until the field is reversed. At some point, after the field changes sign, the dipole will switch from the down to the up state and remain in the up state for the rest of the protocol. On average, there will always be a time lag between the time at which the field changes sign and the time at which the dipole reverses orientation. In other paths the dipole will reverse orientation before the field changes sign, that is, when $H < 0$. These sorts of paths become more and more rare as the ramping speed increases. Finally, in the most general case, the dipole can reverse orientation more than once. The dipole will always start in the down orientation and end in the up orientation with multiple transitions occurring along the path.

The work along a given path is given by Eq. (37),

$$W(\Gamma) = -\int_{-\infty}^{\infty} dt\, \dot{H}(t)\mu(t) = -r\int_{-\infty}^{\infty} dt\, \mu(t) \tag{135}$$

Note that because $E_H(\mu) = E_{-H}(-\mu)$, then $\Delta E = 0$ and $Q(\Gamma) = W(\Gamma)$ so heat and work distributions are identical in this example. Moreover, due to the time-reversal symmetry of the ramping protocol, the work distribution $P(W)$ is identical along the forward and reverse processes. Therefore, we expect that the JE (Eq. (40)) and the CFT (Eq. (41)) are both satisfied with $\Delta F = 0$:

$$\frac{P(W)}{P(-W)} = \exp\left(\frac{W}{T}\right); \quad \left\langle \exp\left(-\frac{W}{T}\right)\right\rangle = 1 \tag{136}$$

The exact computation of $P(W)$ in this simple one-dipole model is already a very arduous task that, to my knowledge, has not yet been exactly solved.* We can, however, consider a limiting case and try to elucidate the properties of the work (heat) distribution. Here we consider the limit of large ramping speed r, where the dipole executes just one transition from the down to the up orientation. A few of these paths are depicted in Fig. 13b. This is also called a first-order Markov process because it only includes transitions that occur in just one direction (from down to up). In this reduced and oversimplified description, a path is fully specified by the value of the field H^* at which the dipole reverses orientation. The work along one of these paths is given by

$$W(\Gamma \equiv H^*) = -\lim_{H_0\to\infty}\int_{-H_0}^{H_0} dH\, \mu(H) = ((H^* + H_0) - (H_0 - H^*))\mu = 2\mu H^* \tag{137}$$

*An exact solution to this problem has been recently accomplished by E. Subrt and P. Chvosta [E. Subrt and P. Chvosta, Exact analysis of work fluctuations in two-level systems, *J. Stat. Mech.* (2007) P09019].

According to the second law, $\langle W \rangle = \langle Q \rangle \geq 0$, which implies that the average switching field is positive, $\langle H^* \rangle \geq 0$ (as expected due to the time lag between the reversal of the field and the reversal of the dipole). The work distribution is just given by the switching field distribution $p(H^*)$. This is a quantity easy to compute. The probability that the dipole is in the down state at field H satisfies a master equation that only includes the death process,

$$\frac{\partial p_{\text{down}}(H)}{\partial H} = \frac{-k_{\text{up}}(H)}{r} p_{\text{down}}(H) \tag{138}$$

This equation can be solved exactly:

$$p_{\text{down}}(H) = \exp\left(-\frac{1}{r}\int_{-\infty}^{H} dH\, k_{\text{up}}(H)\right) \tag{139}$$

where we have inserted the initial condition $p_{\text{down}}(-\infty) = 1$. The integral in the exponent can easily be evaluated using (Eqs. (132) and (133). We get

$$p_{\text{down}}(H) = \left(1 + \exp\left(\frac{2\mu H}{T}\right)\right)^{-Tk_0/2\mu r} \tag{140}$$

The switching field probability distribution $p(H^*)$ is given by $p(H^*) = -(p_{\text{down}})'(H^*)$. From Eq. (137), we get

$$P(W) = \frac{k_0}{4\mu r}\left(1 + \exp\left(\frac{W}{T}\right)\right)^{-Tk_0/2\mu r} \frac{\exp(W/2T)}{\cosh(W/2T)} \tag{141}$$

and from this result we obtain the path entropy,

$$\begin{aligned} S(W) = \log(P(W)) &= -\frac{Tk_0}{2\mu r}\log\left(\exp\left(\frac{W}{T}\right) + 1\right) + \frac{W}{2T} \\ &\quad - \log\left(\cosh\left(\frac{W}{2T}\right)\right) + \text{constant} \end{aligned} \tag{142}$$

It is important to stress that Eq. (141) does not satisfy Eq. (136) except in the limit $r \to \infty$, where this approximation becomes exact. We now compute W^{mp} and $W^{\dagger}$ in the large r limit. We obtain, to the leading order,

$$S'(W^{\text{mp}}) = 0 \to W^{\text{mp}} = T\log\left(\frac{2\mu r}{k_0 T}\right) + \mathcal{O}\left(\frac{1}{r}\right) \tag{143}$$

$$S'(W^{\dagger}) = \frac{1}{T} \to W^{\dagger} = -T\log\left(\frac{2\mu r}{k_0 T}\right) + \mathcal{O}\left(\frac{1}{r}\right) \tag{144}$$

so the symmetry in Eq. (126) (or Eq. (127)) is satisfied to the leading order (yet it can be shown how the $1/r$ corrections appearing in $W^{\rm mp}$ and $W^{\dagger}$ (Eqs. (143) and (144)) are different). We can also compute the leading behavior of the fluctuation-dissipation parameter R (Eq. (42)) by observing that the average work $\langle W \rangle$ is asymptotically equal to the most probable work. The variance of the work, σ_W^2, is found by expanding $S(W)$ around $W^{\rm mp}$:

$$S(W) = S(W^{\rm mp}) + \frac{S''(W^{\rm mp})}{2}(W - W^{\rm mp})^2 + (\text{higher order terms}) \quad (145)$$

$$\sigma_W^2 = -\frac{1}{S''(W^{\rm mp})} \quad (146)$$

A simple computation shows that $\sigma_W^2 = 2T$ and, therefore,

$$R = \frac{\sigma_W^2}{2TW_{\rm diss}} \rightarrow \frac{1}{\log(2\mu r/k_0 T)} \quad (147)$$

so R decays logarithmically to zero. The logarithmic increase of the average work with the ramping speed (Eq. (143)) is just a consequence of the logarithmic increase of the average value of the switching field $\langle H^* \rangle$ with the ramping speed. This result has also been predicted for the dependence of the average breakage force of molecular bonds in single molecule pulling experiments. This phenomenology, related to the technique commonly known as dynamic force spectroscopy, allows one to explore free energy landscapes by varying the pulling speed over several orders of magnitude [131, 132].

2. *A Mean-Field Approach*

We now focus our attention on an analytical method useful for computing work distributions, $P(W)$, in mean-field systems. The method has been introduced in Ref. 117 and developed in full generality by A. Imparato and L. Peliti [133, 134]. This section is a bit technical. The reader not interested in the details can just skip this section and go to Section V.C.

The idea behind the method is the following. We express the probability distribution $P(W)$ as a sum over all paths that start from a given initial state. This sum results in a path integral that can be approximated by its dominant solution or classical path in the large N limit, N being the number of particles. The present approach exploits the fact that, as soon as N becomes moderately large, the contribution to the path integral is very well approximated by the classical path. In addition, the classical path exactly satisfies the FT. Here we limit ourselves to show in a very sketchy way how the method applies to solve

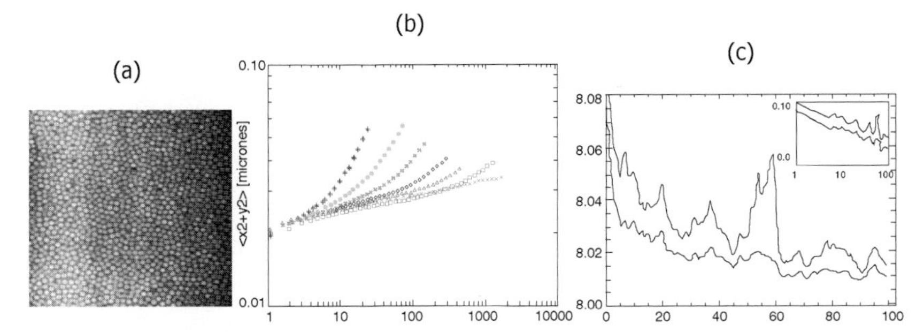

Figure 1. (a) A snapshot picture of a colloidal system obtained with confocal microscope. (b) Aging behavior observed in the mean square displacement, $\langle \Delta x^2 \rangle$, as a function of time for different ages. The colloidal system reorganizes slower as it becomes older. (c) $\gamma = \sqrt{\langle \Delta x^4 \rangle / 3}$ (upper curve) and $\langle \Delta x^2 \rangle$ (lower curve) as a function of the age measured over a fixed time window $\Delta t = 10\,\mathrm{min}$. For a diffusive dynamics both curves should coincide, however these measurements show deviations from diffusive dynamics as well as intermittent behavior. Panels (a) and (b) from http:// www.physics.emory.edu/ weeks/lab/aging.html and Panel (c) from Ref.11.

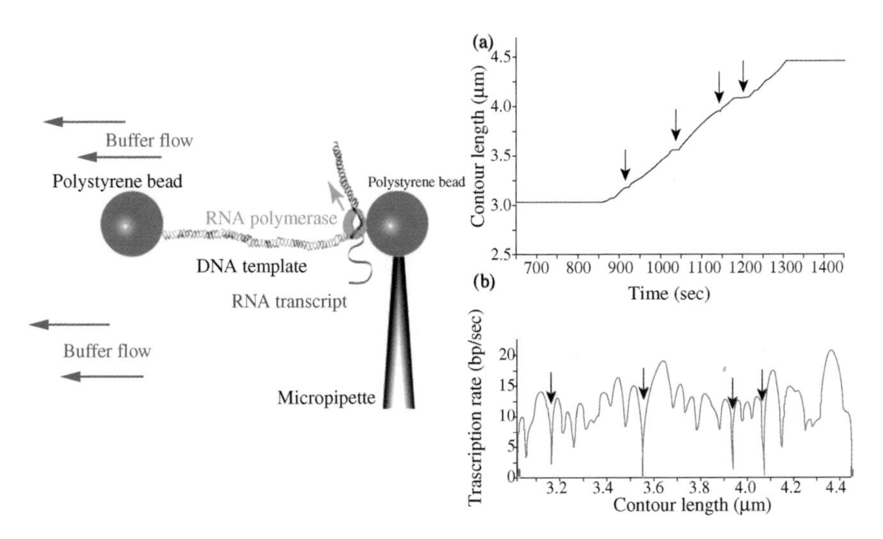

Figure 2. (Left) Experimental setup in force flow measurements. Optical tweezers are used to trap beads but forces are applied on the RNApol–DNA molecular complex using the Stokes drag force acting on the left bead immersed in the flow. In this setup, force assists RNA transcription as the DNA tether between beads increases in length as a function of time. (a) The contour length of the DNA tether as a function of time and (b) the transcription rate as a function of the contour length. Pauses (temporary arrests of transcription) are shown as vertical arrows. (From Ref. 25.)

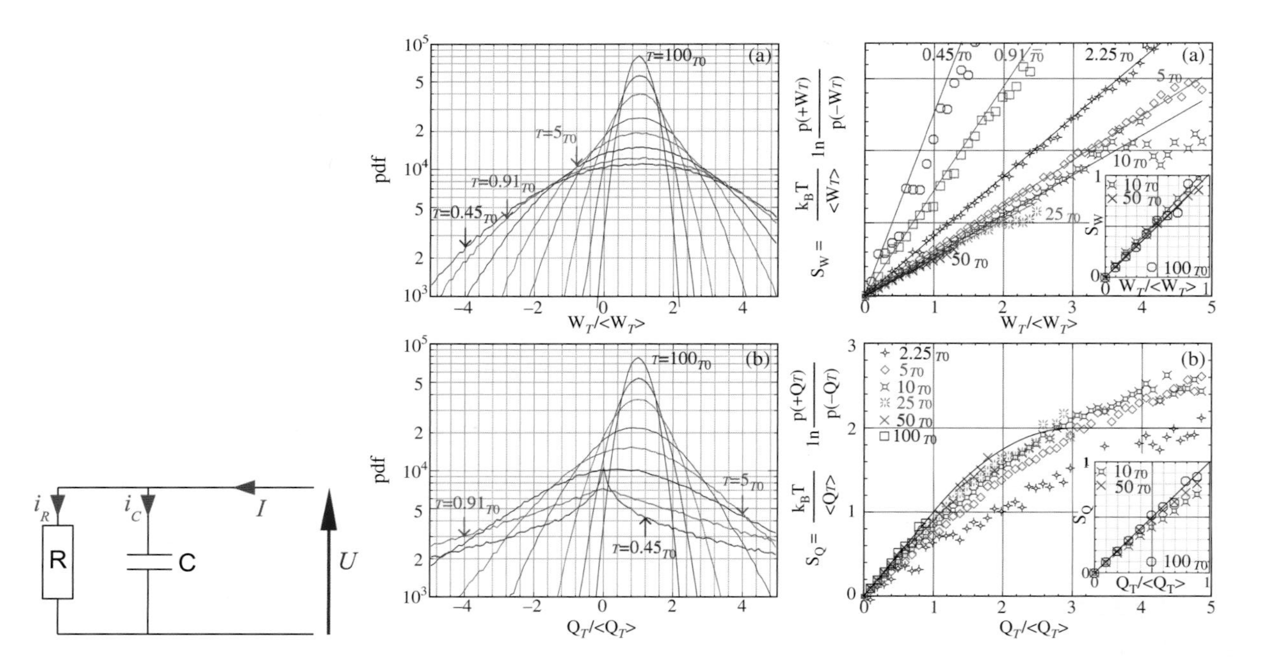

Figure 4. Heat and work fluctuations in an electrical circuit (left). PDF distributions (center) and verification of the FTs (Eqs. (83) and (84)) (right). (From Ref. 68.)

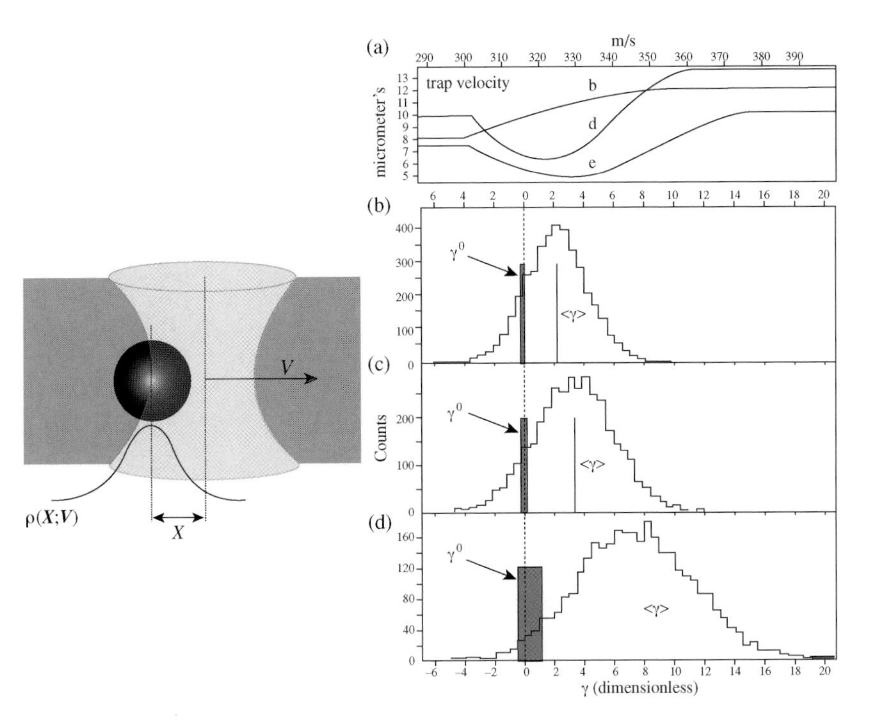

Figure 5. (Left) Bead confined in a moving optical trap. (Right) Total entropy $\mathcal{S}$ distributions (b–d) for the velocity protocols shown in (a). (From Ref. 69.)

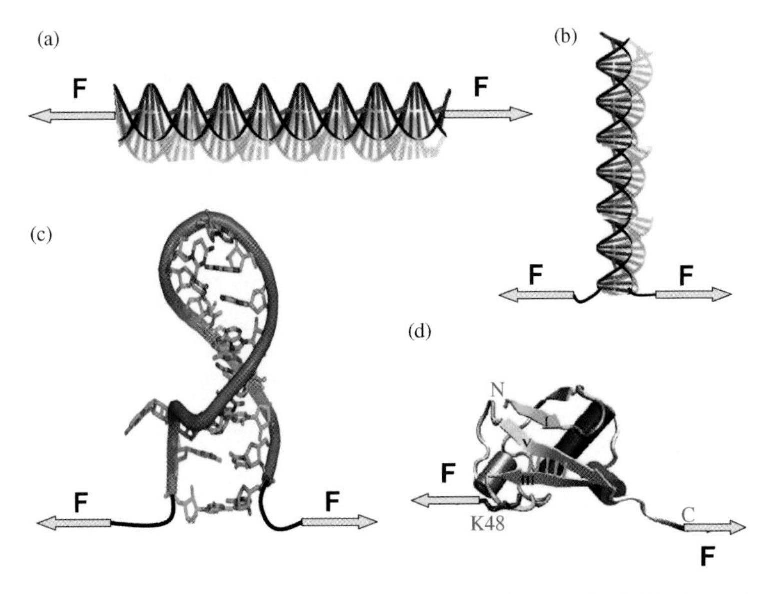

Figure 6. Pulling single molecules. (a) Stretching DNA; (b) unzipping DNA; (c) mechanical unfolding of RNA, and (d) mechanical unfolding of proteins.

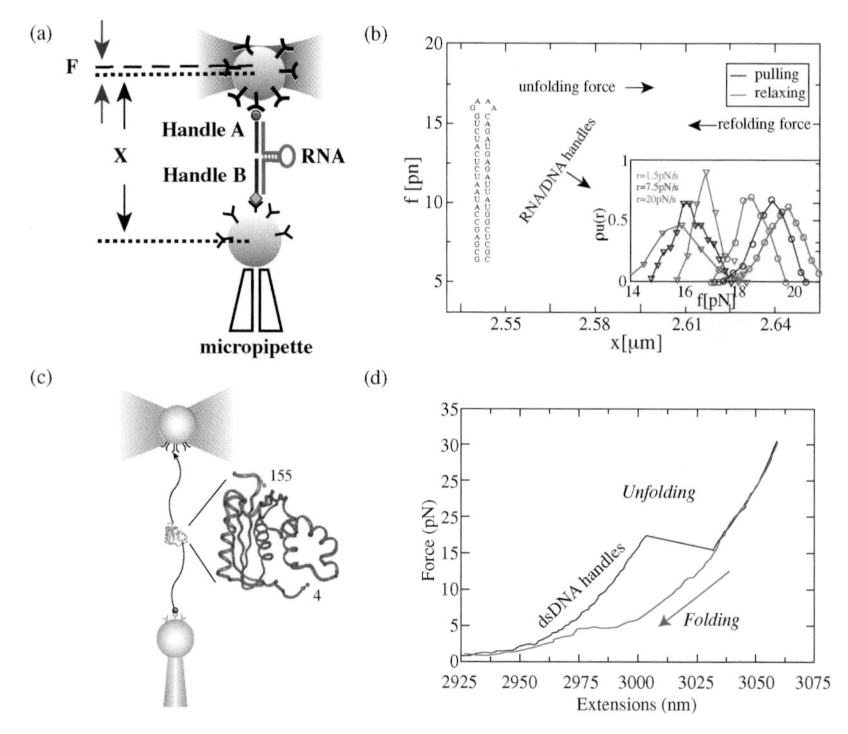

Figure 7. Mechanical unfolding of RNA molecules (a, b) and proteins (c, d) using optical tweezers. (a) Experimental setup in RNA pulling experiments. (b) Pulling cycles in the homologous hairpin and force rip distributions during the unfolding and refolding at three pulling speeds. (c) Equivalent setup in proteins. (d) Force extension curve when pulling the protein RNAseH. Panel (b) is from Ref. 86. Panels (a) and (d) are a courtesy from C. Cecconi [84].

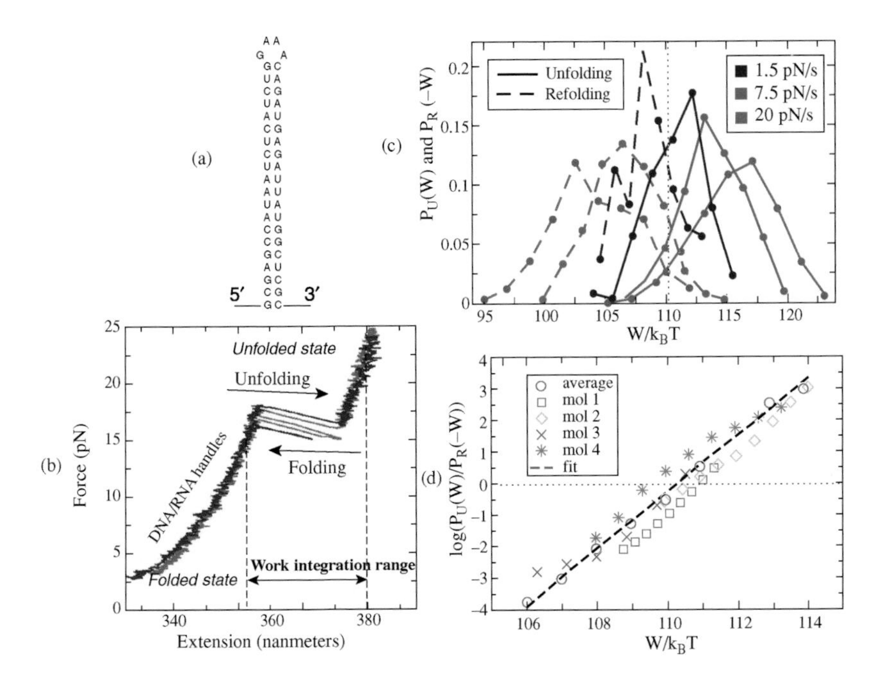

Figure 9. (a) Structure of the homologous CD4 hairpin. (b) FECs at a loading rate of 1.7 pN/s. (c) Unfolding and refolding work distributions at three loading rates (see inset). The unfolding and refolding work distributions cross at a value ΔG independent of the pulling speed as predicted by the CFT. Data correspond to 100,400 and 700 pulls for the lowest, medium, and highest pulling speeds, respectively. (d) Test of the CFT at the intermediate loading rate 7.5 pN/s for four different tethers. The trend of the data is reproducible from tether to tether and consistent with the CFT prediction. (From Ref. 102.)

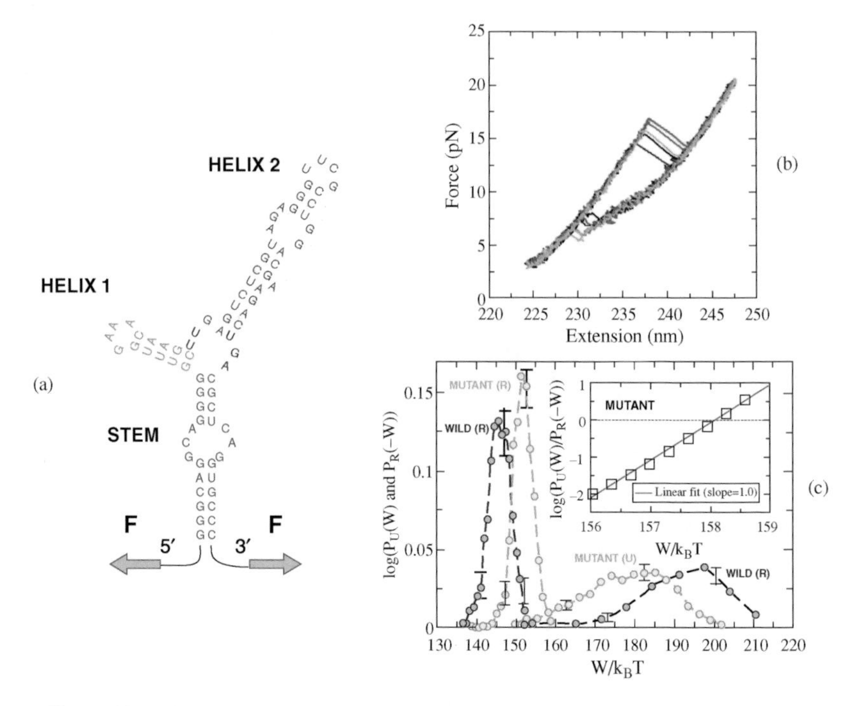

Figure 10. (a) Secondary structure of the three-way junction S15. (b) A few FECs for the wild type. (c) Unfolding/refolding work distributions for the wild type and the mutant. (Inset) Experimental verification of the validity of the CFT for the mutant, where unfolding and refolding distributions overlap each other over a limited range of work values. Data correspond to 900 pulls for the wild type and 1200 pulls for the mutant. (From Ref. 102.)

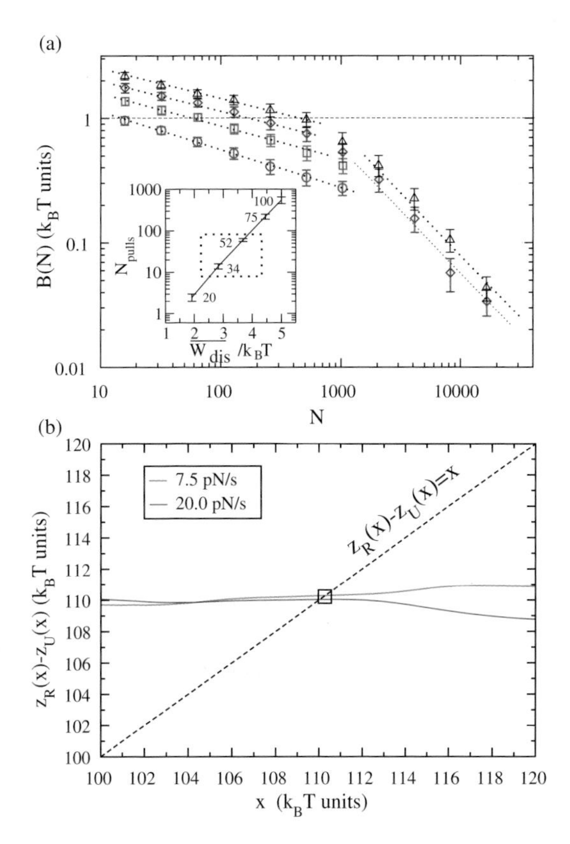

Figure 11. (a) Bias as a function of the number of pulls N for a two-states model. The inset shows the number of pulls as a function of the dissipated work required to recover the free energy with an error within $1k_BT$. (b) Function $z_R - z_F$ for the data shown in Fig. 9c at the two largest pulling speeds. Panel (a) from (Refs. 3 and 101; Panel (b) from the supplementary material in Ref. 102.)

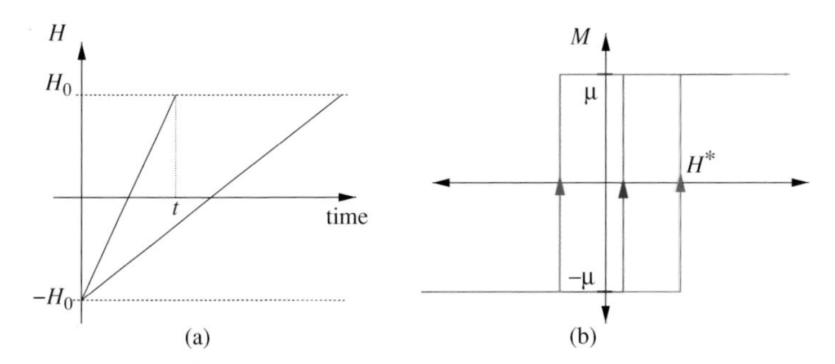

Figure 13. (a) Ramping protocol. The ramping speed is defined by $r = 2H_0/t$ where t is the duration of the ramp. (b) Three examples of paths where the down dipole reverses orientation at different values of the field, H^*.

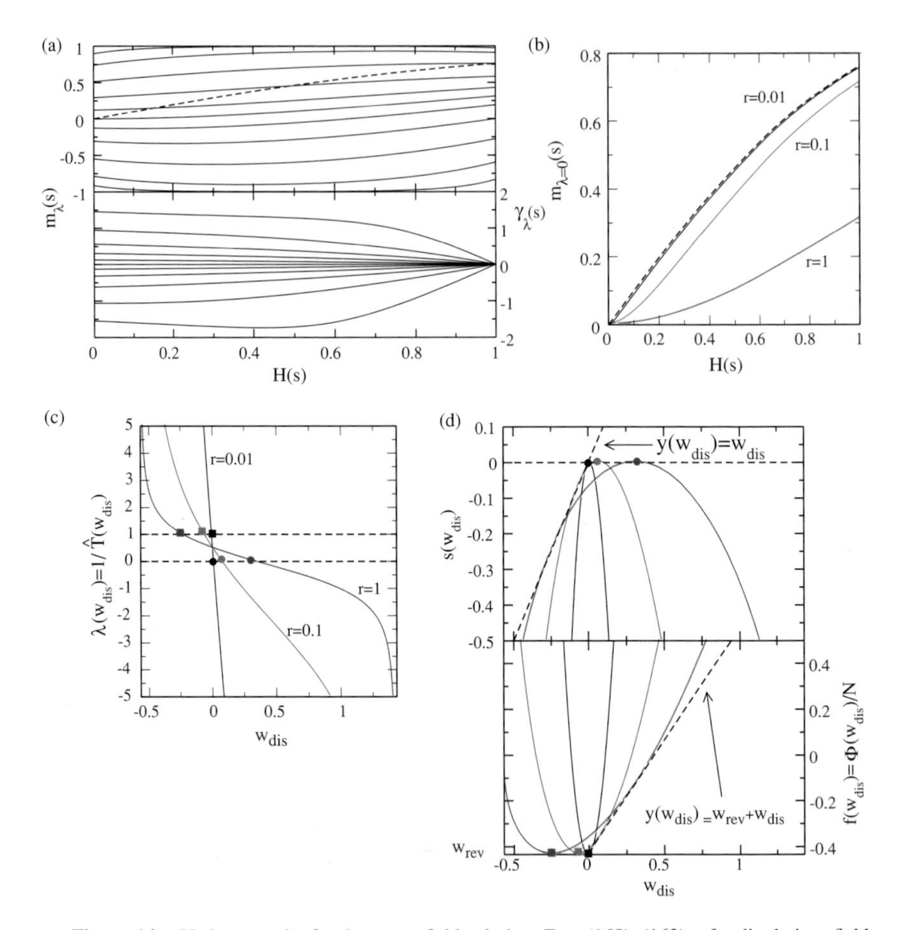

Figure 14. Various results for the mean-field solution, Eqs. (160)–(162), of a dipole in a field that is ramped from $H_i = 0$ to $H_f = 1$. (a) Fields $m_\lambda(s)$ and $\gamma_\lambda(s)$ at the ramping speed $r = 1$. Curves correspond to different values of λ ($\lambda = -5, -2, -1, -0.5, -0.2, 0, 0.2, 0.5, 1, 2, 5$ from top to bottom in the upper and lower panel). The dashed line in $m_\lambda(s)$ is the equilibrium solution $m_{\text{eq}}(H) = \tanh(H)$ corresponding to the reversible process $r \to 0$. (b) Magnetization $m_\lambda(s)$ for the most probable path $\lambda = 0$. The dashed line corresponds to the reversible trajectory, $r \to 0$. (c) Lagrange multiplier $\lambda(w_{\text{dis}})$ for three ramping speeds. The intersection of the different curves with the dashed line $\lambda = 0$ gives w^{mp} (filled circles) whereas the intersection with $\lambda = -1$ gives $w^\dagger$ (filled squares). The intersection of all three curves around $\lambda = 0.5$ is only accidental (looking at a larger resolution such crossing is not seen). (d) Path entropy and free energy corresponding to the solutions shown in (b, c) (larger speeds correspond to wider distributions). Path entropies are maximum and equal to zero at $w_{\text{dis}}^{\text{mp}} = w^{\text{mp}} - w_{\text{rev}}$ (filled circles) whereas path free energies are minimum and equal to $f^\dagger = w_{\text{rev}}$ at $w_{\text{dis}}^\dagger$ (filled squares). (From Ref. 117.)

the specific example shown in Section V.B.1. A detailed and more complete derivation of the method can be found in Refs. 117 and 134.

We come back to the original model, Eqs. (132) and (133), and include all possible paths where the dipole reverses orientation more than once. The problem now gets too complicated, so we modify the original model by considering an ensemble of noninteracting N identical dipoles. A configuration in the system is specified by the N-component vector $\mathcal{C} \equiv \{\vec{\mu} = (\mu_i)_{1 \leq i \leq N}\}$ with $\mu_i = \pm\mu$ the two possible orientations of each dipole. A path is specified by the time sequence $\Gamma \equiv \{\vec{\mu}(s); 0 \leq s \leq t\}$. The energy of the system is given by

$$E(\mathcal{C}) = -hM(\mathcal{C}) = -h \sum_{i=1}^{N} \mu_i \tag{148}$$

where $M = \sum_i \mu_i$ is the total magnetization. The equilibrium free energy is $F = -N\log(2\cosh(\mu H/T))$ and the kinetic rules are the same as given in Eqs. (132) and (133) and are identical for each dipole. The work along a given path is given by Eq. (135),

$$W(\Gamma) = -\int_0^t ds\, \dot{H}(s) M(s) \tag{149}$$

so the work probability distribution is given by the path integral,

$$P(W) = \sum_{\Gamma} P(\Gamma)\delta(W(\Gamma) - W) = \int \mathcal{D}[\vec{\mu}]\delta\left(W + \int_0^t ds\, \dot{H}(s)M(s)\right) \tag{150}$$

where we have to integrate over all paths where $\vec{\mu}$ starts at time 0 in a given equilibrium state up to a final time t. To solve Eq. (150) we use the integral representation of the delta function,

$$\delta(x) = (1/2\pi) \int_{-\infty}^{\infty} d\lambda \, \exp(-i\lambda x) \tag{151}$$

We also insert the following factor,

$$1 = \int \frac{\mathcal{D}[\gamma]\mathcal{D}[m]}{2\pi} \exp\left(\frac{i}{\Delta t} \int \gamma(s) \left(m(s) - \frac{1}{N}\sum_{i=1}^{N} \mu_i(s)\right)\right) \tag{152}$$

where Δt is the discretization time step and we have introduced new scalar fields $\gamma(s)$ and $m(s)$. After some manipulations one gets a closed expression for the work distribution $P(w)$ ($w = W/N$ is the work per dipole). We quote the final result [117]:

$$P(w) = \mathcal{N} \int d\lambda \mathcal{D}[\gamma]\mathcal{D}[m] \exp(Na(w, \lambda, \gamma, m)) \tag{153}$$

where $\mathcal{N}$ is a normalization constant and a represents an action given by

$$a(w, \lambda, \gamma, m) = \lambda\left(w + \int_0^t ds\dot{H}(s)M(s)\right) \tag{154}$$

$$+\frac{1}{2}\int_0^t ds(m(s)(2\dot{\gamma}(s) + c(s)) + d(s))$$

$$+ \log\big(\exp(\gamma(0))k_{\text{up}}(H_i) + \exp(-\gamma(0)k_{\text{down}}(H_i))\big) \tag{155}$$

with

$$c(s) = k_{\text{down}}(H(s))(\exp(-2\gamma(s)) - 1) - k_{\text{up}}(H(s))(\exp(2\gamma(s)) - 1) \tag{156}$$

$$d(s) = k_{\text{down}}(H(s))(\exp(-2\gamma(s)) - 1) + k_{\text{up}}(H(s))(\exp(2\gamma(s)) - 1) \tag{157}$$

where the rates k_{up} and k_{down} are given in Eqs. (132) and (133) and we have assumed an initial equilibrium state at the the initial value of the field, $H(0) = H_i$. Equation (155) has to be solved together with the boundary conditions:

$$\gamma(t) = 0; \quad m(0) = \tanh\left(\gamma(0) + \frac{\mu H_i}{T}\right) \tag{158}$$

Note that these boundary conditions break causality. The function γ has the boundary at the final time t whereas m has the boundary at the initial time 0. Causality is broken because by imposing a fixed value of the work w along the paths we are constraining the time evolution of the system.

To compute $P(w)$ we take the large volume limit $N \to \infty$ in Eq. (153). For a given value of w the probability distribution is given by

$$P(w) \propto \exp(Ns(w)) = \exp(Na(w), \lambda(w), \gamma_w(s), m_w(s)) \tag{159}$$

where s is the path entropy (Eq. (119)) and the functions $\lambda(w), \gamma_w(s)$ and $m_w(s)$ are solutions of the saddle point equations,

$$\frac{\delta a}{\delta \lambda} = w + \mu \int_0^t m_w(s)\dot{H}(s)\, ds = 0 \tag{160}$$

$$\frac{\delta a}{\delta \gamma(s)} = \dot{m}_w(s) + m_w(s)(k_{\text{up}}(s) + k_{\text{down}}(s))$$

$$- (k_{\text{up}}(s) - k_{\text{down}}(s)) + m_w(s)d_w(s) + c_w(s) = 0 \tag{161}$$

$$\frac{\delta a}{\delta m(s)} = \dot{\gamma}_w(s) + \lambda(w)\mu\dot{H}(s) + \frac{1}{2}c_w(s) = 0 \tag{162}$$

These equations must be solved together with the boundary conditions in Eq. (158). Note that we use the subindex (or the argument) w in all fields (λ, m, γ) to emphasize that there exists a solution of these fields for each value of the work w. From the entropy s in Eq. (159) we can evaluate the path free energy, the path temperature, and the values W^{mp} and $W^{\dagger}$ introduced in Section V.A. We enumerate the different results.

- **The Path Entropy s(w).** By inserting Eq. (162) into Eq. (155), we get

$$s(w) = \lambda(w)w + \frac{1}{2}\int_0^t d_w(s)\,ds + \log\big(\exp(\gamma(0))k_{\mathrm{up}}(H_i) + \exp(-\gamma(0))k_{\mathrm{down}}(H_i)\big) \tag{163}$$

From the stationary conditions—Eqs. (160)–(162)—the path entropy in Eq. (159) satisfies

$$s'(w) = \frac{ds(w)}{dw} = \frac{\partial a(w, \lambda(w), \gamma_w(s), m_w(s))}{\partial w} = \lambda(w) \tag{164}$$

The most probable work can be determined by finding the extremum of the path entropy $s(w)$,

$$s'(w^{\mathrm{mp}}) = \lambda(w^{\mathrm{mp}}) = 0 \tag{165}$$

where we used Eq. (123). The saddle point equations (160)–(162) give $\gamma_{w^{\mathrm{mp}}}(s) = c_{w^{\mathrm{mp}}}(s) = d_{w^{\mathrm{mp}}}(s) = 0$ and

$$\dot{m}_{w^{\mathrm{mp}}}(s) = -m_{w^{\mathrm{mp}}}(s)(k_{\mathrm{up}}(s) + k_{\mathrm{down}}(s)) + (k_{\mathrm{up}}(s) - k_{\mathrm{down}}(s)) \tag{166}$$

which is the solution of the master equation for the magnetization. The stationary solution of this equation gives the equilibrium solution $m^{\mathrm{eq}}(s) = \tanh(\mu H(s)/T)$ corresponding to a quasistationary or reversible process.

- **The Path Free Energy.** The path free energy $f = \Phi/N$ (Eq. (121)) is given by

$$f^{\dagger} = f(w^{\dagger}) = \frac{\Delta F}{N} = w_{\mathrm{rev}} = w^{\dagger} - Ts(w^{\dagger}) = \frac{T}{2}\int_0^t d_{w^{\dagger}}(s)\,ds \tag{167}$$

where $w^{\dagger}$ is given by

$$s'(w^{\dagger}) = \lambda(w^{\dagger}) = \frac{1}{T} \tag{168}$$

and the path temperature (Eq. (124)) satisfies the identity

$$\hat{T}(w) = \frac{1}{\lambda(w)}; \quad \hat{T}(w^{\dagger}) = T \tag{169}$$

This set of equations can be solved numerically. Figure 14 shows some of the results.

C. Large Deviation Functions and Tails

A large deviation function $\hat{P}(x)$ of a function $P_L(x)$ is defined if the following limit exists:

$$\hat{P}(x) = \lim_{L\to\infty} \frac{1}{L} \log\left(P_L\left(\frac{x}{L}\right)\right) \tag{170}$$

From this point of view, the distribution of the entropy production in a NESS, $P(a)$ (Eq. (55)), where $a = \mathcal{S}_\mathrm{p}/\langle\mathcal{S}_\mathrm{p}\rangle$, and the work distribution $P(W)$ (Eq. (159)), define large deviation functions. In the first case, $\lim_{t\to\infty} f_t(a)$ is the large deviation function (e.g. Eq. (84)), the average entropy production $\langle\mathcal{S}_\mathrm{p}\rangle$ being the equivalent of L in Eq. (170). In the second case, the path entropy $s(w) = S(W)/N$ (Eqs. (119) and (159)) is a large deviation function, where L in Eq. (170) corresponds to the size N. Large deviation functions are interesting for several reasons.

- **Nonequilibrium Theory Extensions.** By knowing the large deviation function of an observable (e.g., the velocity or position density) in a nonequilibrium system, we can characterize the probability of macroscopic fluctuations. For example, by knowing the function $s(w)$ we can determine the probability of macroscopic work fluctuations $\delta W \propto N$, where N is the size of the system. Large deviations (e.g., in work) may depend on the particular details (e.g., the rules) of the nonequilibrium dynamics. In contrast, small deviations (i.e., $\delta W \propto \sqrt{N}$) are usually insensitive to the microscopic details of the dynamics. Nonequilibrium systems are nonuniversal and often strongly dependent on the microscopic details of the system. In this regard, understanding large or macroscopic deviations may be a first step in establishing a general theory for nonequilibrium systems.
- **Spectrum of Large Deviations.** There are few examples where large deviations can be analytically solved. Over the past years a large amount of work has been devoted to understanding large deviations in some statistical models such as exclusion processes. General results include the additivity principle in spatially extended systems [135–137] and the existence of exponential tails in the distributions [138]. These general results and the spectrum of large deviations are partially determined by the validity of the FT (Eq. (27)), which imposes a specific relation between the forward and the reverse work/heat distributions. For example, exponential tails in the work distribution $P(W)$ (Eq. (119)) correspond to

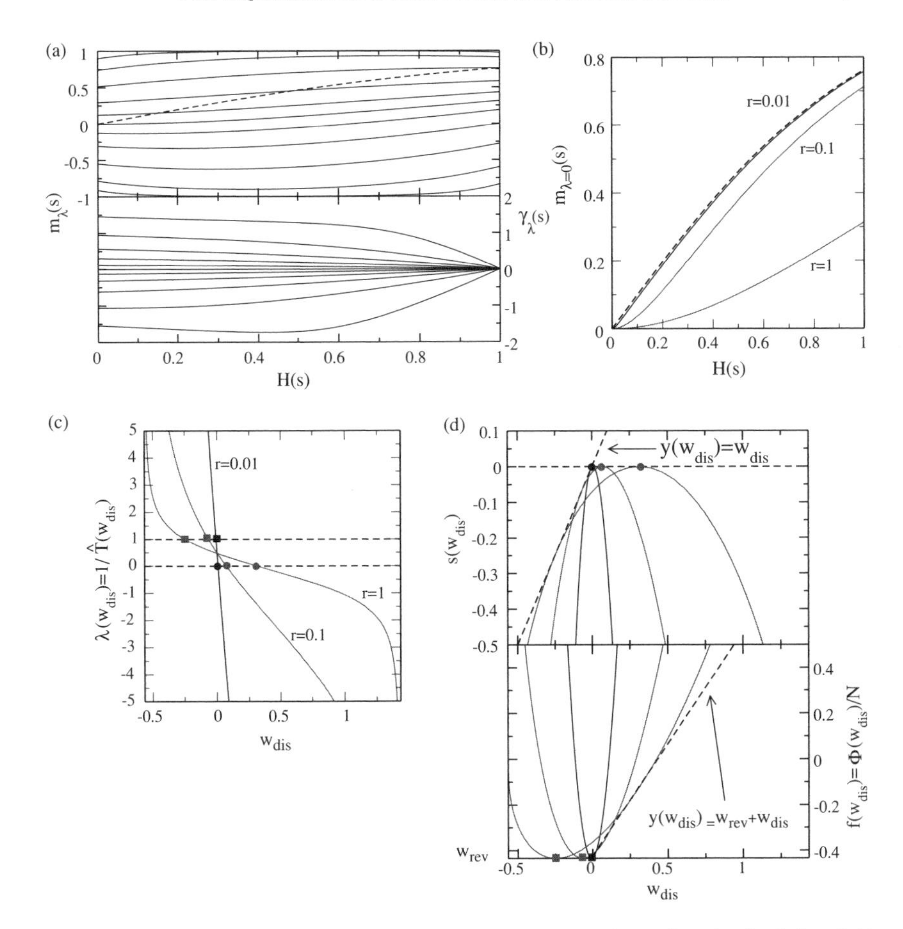

Figure 14. Various results for the mean-field solution, Eqs. (160)–(162), of a dipole in a field that is ramped from $H_i = 0$ to $H_f = 1$. (a) Fields $m_\lambda(s)$ and $\gamma_\lambda(s)$ at the ramping speed $r = 1$. Curves correspond to different values of λ ($\lambda = -5, -2, -1, -0.5, -0.2, 0, 0.2, 0.5, 1, 2, 5$ from top to bottom in the upper and lower panel). The dashed line in $m_\lambda(s)$ is the equilibrium solution $m_{eq}(H) = \tanh(H)$ corresponding to the reversible process $r \to 0$. (b) Magnetization $m_\lambda(s)$ for the most probable path $\lambda = 0$. The dashed line corresponds to the reversible trajectory, $r \to 0$. (c) Lagrange multiplier $\lambda(w_{dis})$ for three ramping speeds. The intersection of the different curves with the dashed line $\lambda = 0$ gives w^{mp} (filled circles) whereas the intersection with $\lambda = -1$ gives $w^\dagger$ (filled squares). The intersection of all three curves around $\lambda = 0.5$ is only accidental (looking at a larger resolution such crossing is not seen). (d) Path entropy and free energy corresponding to the solutions shown in (b, c) (larger speeds correspond to wider distributions). Path entropies are maximum and equal to zero at $w_{dis}^{mp} = w^{mp} - w_{rev}$ (filled circles) whereas path free energies are minimum and equal to $f^\dagger = w_{rev}$ at $w_{dis}^\dagger$ (filled squares). (From Ref. 117.) (See color insert.)

a path entropy $S(W)$ that is linear in W. This is the most natural solution of the FT; see Eq. (125).

- **Physical Interpretation of Large Deviations.** In small systems, large deviations are common and have to be considered as important as small deviations. This means that, in order to understand the nonequilibrium behavior of small systems, a full treatment of small and large deviations may be necessary. The latter are described by the shape of the large deviation function. The physical interpretation of small and large deviations may be different. For example, if we think of the case of molecular motors, small deviations (with respect to the average) of the number of mechanochemical cycles may be responsible for the average speed of a molecular motor, whereas large deviations may be relevant to understanding why molecular motors operate so efficiently along the mechanochemical cycles.

1. *Work and Heat Tails*

Let us consider the case of a NETS that starts initially in equilibrium and is driven out of equilibrium by some external driving forces. As we have seen in Eq. (159), $(1/N)\log(P(w)) = s(w)$ is a large deviation function. At the same time we could also consider the heat distribution $P(Q)$ and evaluate its large deviation function $(1/N)\log(P(Q)) = s(q)$, where $q = Q/N$. Do we expect $s(q)$ and $s(w)$ to be identical? Heat and work differ by a boundary term, the energy difference. Yet the energy difference is extensive with N; therefore, boundary terms modify the large deviation function so we expect that $s(q)$ and $s(w)$ are different. An interesting example is the case of the bead in the harmonic trap discussed in Section IV.A. Whereas the work distribution measured along arbitrary time intervals is always Gaussian, the heat distribution is characterized by a Gaussian distribution for small fluctuations $\delta Q = Q - \langle Q \rangle \propto \sqrt{t}$, plus exponential tails for large deviations $\delta Q \propto t$. The difference between the large deviation function for the heat and the work arises from a boundary term, the energy difference. Again, in the large t limit, the boundary term is important for large fluctuations when $a = |Q|/\langle Q \rangle \geq a^* = 1$ (Eq. (84)). Large deviation functions always depend on boundary terms and these can never be neglected.

Let us come back now to the example of Section V.B.1, where we considered work distributions in a system of noninteracting dipoles driven by an externally varying magnetic field. Again, we will focus the discussion on the particular case where the initial value of the field is negative and large, $H_i = -H_0 \to -\infty$, and the field is ramped at speed r until reaching the final value $H_f = H_0 \to \infty$. In this case, $Q = W$ for individual paths so both large deviation functions $s(q), s(w)$ are identical. In what follows we will use heat instead of work for the arguments of all functions. In addition $s_{\rm F} = s_{\rm R}$ due to the time-reversal symmetry of the protocol. Exponential tails are indicated by a path temperature $\hat{T}(q)$ (Eq. (124)), which is constant along a finite interval of heat values.

In Section V.B.1 we have evaluated the path entropy $s(q)$ (Eq. (163)) for an individual dipole ($N = 1$) in the approximation of a first-order Markov process. The following result has been obtained (Eq. (142)):

$$s(q) = -\frac{Tk_0}{2\mu r}\log\left(\exp\left(\frac{q}{T}\right) + 1\right) + \frac{q}{2T} - \log\left(\cosh\left(\frac{q}{2T}\right)\right) + \text{constant} \quad (171)$$

For $|q| \to \infty$, we get

$$s(q \to \infty) = -\frac{qk_0}{2\mu r} + \mathcal{O}\left(\exp\left(-\frac{q}{T}\right)\right) \quad (172)$$

$$s(q \to -\infty) = \frac{q}{T} + \mathcal{O}\left(\exp\left(\frac{q}{T}\right)\right) \quad (173)$$

The linear dependence of $s(q)$ on q leads to

$$\hat{T}(q \to \infty) = T^- = -\frac{2\mu r}{k_0} \quad (174)$$

$$\hat{T}(q \to -\infty) = T^+ = T \quad (175)$$

$$(176)$$

where we use the notation T^+ and T^- to stress the fact that these path temperatures are positive and negative, respectively. Both path temperatures are constant and lead to exponential tails for positive and negative work values. Note that Eq. (125) reads

$$s(q) - s(-q) = \frac{q}{T} \to (s)'(q) + (s)'(-q) = \frac{1}{T} \to \frac{1}{T^+} + \frac{1}{T^-} = \frac{1}{T} \quad (177)$$

which is satisfied by (Eqs. (174) and (175)) up to $1/r$ corrections.

Another interesting limit is the quasistatic limit $r \to 0$. Based on the numerical solution of the saddle point equations (160)–(162), it was suggested in Ref. 117 that $\hat{T}(q)$ converged to a constant value over a finite range of work values. Figure 15a shows the results obtained for the heat distributions, whereas the path temperature is shown in Fig. 15b. A more detailed analysis [134] has shown that a plateau is never fully reached for a finite interval of heat values when $r \to 0$. The presence of a plateau has been interpreted as the occurrence of a first-order phase transition in the path entropy $s(q)$ [134].

An analogy between the different type of work/heat fluctuations and the emission of light radiation by atoms in a cavity can be established. Atoms can absorb and reemit photons following two different mechanisms. One type of

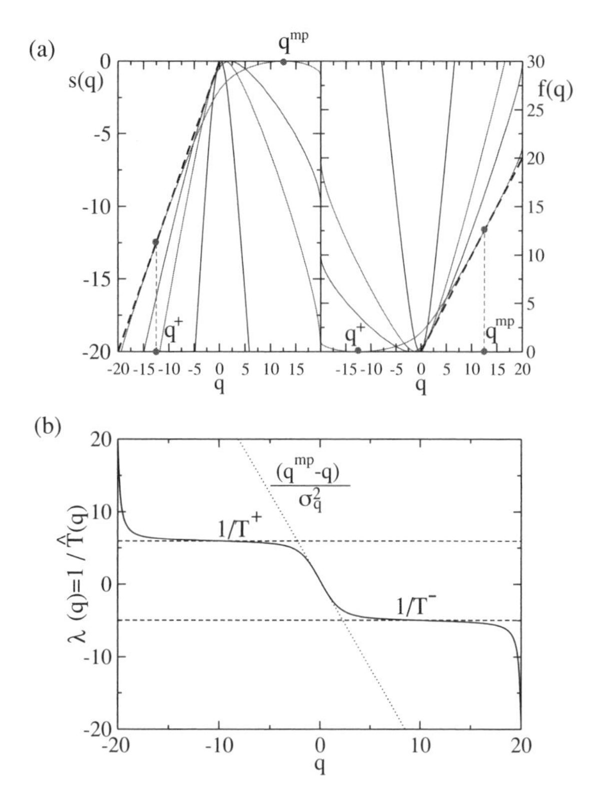

Figure 15. (a) Heat distributions (path entropy $s(q)$ and path free energy $f(q)$) evaluated at four ramping speeds $r = 0.1, 0.5, 1, 10$ (from the narrowest to the widest distributions). The dashed line in the left panel is $y(q) = q/T$ (we take $T = 1$) and is tangent to $s(q)$ at $q^{\dagger}$ (dots are shown for $r = 10$). The dashed line in the right panel corresponds to $y(q) = q$ and is tangent to the function $f(q)$ at the value q^{mp} (dots shown for $r = 10$). (b) $\lambda(q)$ for the lowest speed $r = 0.1$. It shows a linear behavior for small values of q, $\lambda(q) = (1/\sigma_q^2)(q^{\text{mp}} - q)$ and two plateaus for $q \gg 1$ and $q \ll -1$. The former contributes as a Gaussian component to the heat distribution describing the statistics of small deviations with respect to the most probable value (stimulated sector). The latter gives rise to two exponential tails for the distribution describing the statistics of rare events (spontaneous sector). (Adapted from Ref. 117.)

radiative mechanism is called stimulated because it depends on the density of blackbody radiation in the cavity (directly related to the temperature of the cavity). The other radiative mechanism is called spontaneous and is independent of the density of radiation in the cavity (i.e., it does not depend on its temperature). The stimulated process contributes to the absorption and emission of radiation by atoms. The spontaneous process only contributes to the emitted radiation. In general, the path entropy $s(w)$ contains two sectors reminiscent of the stimulated and spontaneous processes in the blackbody radiation.

- **The FDT or Stimulated Sector.** This sector is described by Gaussian work fluctuations (Eq. (128)) leading to $s(q) = -(q - q^{\mathrm{mp}})^2/(2\sigma_q^2)$ +constant. Therefore, we get Eq. (129),

$$\lambda(q) = \frac{1}{\hat{T}(q)} = -\frac{q - q^{\mathrm{mp}}}{\sigma_q^2} \tag{178}$$

which behaves linearly in q for small deviations around q^{mp}. Note that $\hat{T}(q)$ satisfies Eq. (177) and, therefore,

$$\sigma_q^2 = 2Tq^{\mathrm{mp}} \tag{179}$$

leading to a fluctuation-dissipation parameter $R = 1$, a result equivalent to the validity of the fluctuation-dissipation theorem (FDT). This sector we call stimulated because work fluctuations (Eq. (179)) depend directly on the temperature of the bath.

- **The Large Deviation or Spontaneous Sector.** Under some conditions this sector is well reproduced by exponential work tails describing large or macroscopic deviations. In this sector,

$$\frac{1}{\hat{T}(q)} - \frac{1}{\hat{T}(-q)} = \frac{1}{T} \to \frac{1}{T^+} + \frac{1}{T^-} = \frac{1}{T} \tag{180}$$

The physical interpretation of T^+ and T^- is as follows. Because T^- is negative, T^- describes fluctuations where net heat is released to the bath, whereas T^+ is positive and describes fluctuations where net heat is absorbed from the bath. Equation (180) imposes $T^+ < |T^-|$, implying that large deviations also satisfy the second law: the average net amount of heat supplied to the bath ($\propto |T^-|$) is always larger than the average net heat absorbed from the bath ($\propto |T^+|$). In the previous example Eqs. (174) and (175), T^+ converges to the bath temperature whereas T^- diverges to $-\infty$ when $r \to \infty$. We call this sector spontaneous because the energy fluctuations mainly depend on the nonequilibrium protocol (in the current example, such dependence is contained in the r dependence of T^-, Eq. (174)).

2. *The Bias as a Large Deviation Function*

The bias defined in Eq. (103) is still another example of a large deviation function. Let us define the variable

$$X = \sum_{i=1}^{N} \exp\left(-\frac{W_i}{T}\right) \tag{181}$$

where $W_i \to W_i - \Delta F$ stands for the dissipated work. The free energy estimate in Eq. (103) satisfies the relation

$$x = \exp\left(-\frac{F^{\mathrm{JE}} - \Delta F}{T}\right); \quad x = \frac{X}{N} = \frac{1}{N}\sum_{i=1}^{N} \exp(-W_i) \tag{182}$$

where N is the total number of experiments. The N values of W_i are extracted from a distribution $P(W)$ that satisfies the relations

$$\langle 1 \rangle = \int_{-\infty}^{\infty} P(W)dW = 1; \; \langle \exp(-W) \rangle = \int_{-\infty}^{\infty} \exp(-W)P(W)dW = 1 \tag{183}$$

We follow the same procedure as in Section IV.B.2 and extract N different values of W_i to obtain a single x using Eq. (182). By repeating this procedure a large number of times, M, we generate the probability distribution of x, which we will call $\mathcal{P}_N(x)$, in the limit $M \to \infty$. The bias in Eq. (104) is defined by

$$B(N) = -T\langle \log(x) \rangle = -T\int_{-\infty}^{\infty} \log(x)\mathcal{P}_N(x)\, dx \tag{184}$$

In the following we show that $\mathcal{P}_N(x)$ defines a large deviation function in the limit $N \to \infty$. We write

$$\begin{aligned}
\mathcal{P}_N(x) &= \int \prod_{i=1}^{N} dW_i P(W_i) \delta\left(x - \frac{1}{N}\sum_{i=1}^{N} \exp(-W_i)\right) \\
&= \frac{1}{2\pi i}\int_{-i\infty}^{i\infty} d\mu \exp\left(\mu x - \frac{\mu}{N}\sum_{i=1}^{N} \exp(-W_i)\right)\prod_{i=1}^{N} P(W_i)dW_i \\
&= \frac{N}{2\pi i}\int_{-i\infty}^{i\infty} d\hat{\mu} \exp\left(N\hat{\mu}x + N\log\left(\int dWP(W)\exp(-\hat{\mu}\exp(-W))\right)\right) \\
&= \frac{N}{2\pi i}\int_{-i\infty}^{i\infty} d\hat{\mu} \exp(Ng(\hat{\mu},x)) \approx_{N\to\infty} \exp(Ng(\hat{\mu}^*,x))
\end{aligned} \tag{185}$$

where in the second line we used the integral representation of the delta function (Eq. (151)); in the third line we separate the integrals and independently integrate the contribution of each variable W_i; in the last line we apply the saddle point integration method to the function $g(\hat{\mu}, x)$ defined as

$$g(\hat{\mu}, x) = \hat{\mu}x + \log\left(\int_{-\infty}^{\infty} \exp(-\hat{\mu}\exp(-W))\right) \tag{186}$$

where $\hat{\mu}^*$ is equal to the absolute maximum of $g(\hat{\mu}, x)$,

$$\left(\frac{\partial g(\hat{\mu}, x)}{\partial \hat{\mu}}\right)_{\hat{\mu}=\hat{\mu}^*} = 0 \rightarrow x = \ll \exp(-W) \gg_{\hat{\mu}^*} \tag{187}$$

with

$$\ll \cdots \gg_{\hat{\mu}} = \frac{\int_{-\infty}^{\infty} \exp(-W)\exp(-\hat{\mu}\exp(-W))P(W)\,dW}{\int_{-\infty}^{\infty} \exp(-\hat{\mu}\exp(-W))P(W)\,dW} \tag{188}$$

The function $g(\hat{\mu}, x)$ evaluated at $\hat{\mu} = \hat{\mu}^*$ defines a large deviation function (Eq. (170)):

$$g^*(x) = g(\hat{\mu}^*(x), x) = \lim_{N\to\infty} \frac{1}{N}\log\left(\mathcal{P}_N\left(\frac{X}{N}\right)\right) = \lim_{N\to\infty}\frac{1}{N}\log(\mathcal{P}_N(x)) \tag{189}$$

Using Eq. (189), we can write for the bias in Eq. (184) in the large N limit

$$B(N) = -T\frac{\int dx \log(x)\exp(Ng^*(x))}{\int dx\, \exp(Ng^*(x))} \tag{190}$$

The integrals in the numerator and denominator can be estimated by using the saddle point method again. By expanding $g^*(x)$ around the maximum contribution at $x^{\max}$, we get, up to second order,

$$g^*(x) = g^*(x^{\max}) + \tfrac{1}{2}(g^*)''(x^{\max})(x - x^{\max})^2 \tag{191}$$

To determine $x^{\max}$, we compute first

$$\begin{aligned}(g^*)'(x) &= \left(\frac{\partial g(\hat{\mu}, x)}{\partial \hat{\mu}}\right)_{\hat{\mu}=\hat{\mu}^*(x)}\left(\frac{d\hat{\mu}^*(x)}{dx}\right) + \left(\frac{\partial g(\hat{\mu}^*, x)}{\partial x}\right)\\ &= \left(\frac{\partial g(\hat{\mu}^*, x)}{\partial x}\right) = \hat{\mu}^*(x)\end{aligned} \tag{192}$$

where we have used Eqs. (186) and (187). The value $x^{\max}$ satisfies

$$(g^*)'(x^{\max}) = \hat{\mu}^*(x^{\max}) = 0 \tag{193}$$

Inspection of Eqs. (187) and (188) shows that $x^{\max} = 1$. The second term on the rhs of Eq. (191) is then given by

$$(g^*)''(x=1) = (\hat{\mu}^*)'(x=1) = \frac{1}{1 - \langle\exp(-2W)\rangle} \tag{194}$$

where $\langle\cdots\rangle$ denotes the average over the distribution $P(W)$ (Eq. (183)). Using Eq. (194) and inserting Eq. (191) into Eq. (190), we finally obtain

$$B(N) = T\frac{\langle\exp(-2W)\rangle - 1}{2N} + \mathcal{O}\left(\frac{1}{N^2}\right) \tag{195}$$

For a Gaussian distribution, we get $B(N) = T\exp(\sigma_W^2 - 1)/(2N)$. Equation (195) was derived in Ref. 104. For intermediate values of N (i.e., for values of N where $B(N) > 1$), other approaches are necessary.

VI. GLASSY DYNAMICS

Understanding glassy systems (see Section II.A) is a major goal in modern condensed matter physics [139–142]. Glasses represent an intermediate state of matter sharing some properties of solids and liquids. Glasses are produced by fast cooling of a liquid when the crystallization transition is avoided and the liquid enters the metastable supercooled region. The relaxation of the glass to the supercooled state proceeds by reorganization of molecular clusters inside the liquid, a process that is thermally activated and strongly dependent on the temperature. The relaxation of the supercooled liquid is a nonequilibrium process that can be extremely slow leading to aging. The glass analogy is very fruitful to describe the nonequilibrium behavior of a large variety of systems in condensed matter physics, all of them showing a related phenomenology.

The nonequilibrium aging state (NEAS, see Section III.A) is a nonstationary state characterized by slow relaxation and a very low rate of energy dissipation to the surroundings. Aging systems fail to reach equilibrium unless one waits an exceedingly large amount of time. For this reason, the NEAS is very different from either the nonequilibrium transient state (NETS) or the nonequilibrium steady state (NESS).

What do aging systems have in common with the nonequilibrium behavior of small systems? Relaxation in aging systems is driven by fluctuations of a small number of molecules that relax by releasing a small amount of stress energy to the surroundings. These molecules are grouped into clusters often called cooperatively rearranging regions (CRRs). A few observations support this interpretation.

- **Experimental Facts.** Traditionally, the glass transition has been studied with bulk methods such as calorimetry or light scattering. These measurements perform an average over all mesoscopic regions in the sample but are not suitable to follow the motion of individual clusters of a few nanometers in extension. The few direct evidences we have on aging as driven by the rearrangement of small regions comes from AFM measurements on glass surfaces, confocal microscopy of colloids, and the direct observation of molecular motion (NMR and photobleaching tests) [8]. More indirect evidence is obtained from the heterogeneous character of the dynamics, that is, the presence of different regions in the system that show a great disparity of relaxation times [143]. The observation of strong intermittent signals [144] in Nyquist noise measurements while the system ages has been interpreted as the result of CRRs, that is, events corresponding to the rearrangement of molecular clusters. Finally, the direct measure of a correlation length in colloidal glasses hints at the existence of CRRs [145]. Future accomplishments in this area are expected to come from developments in micromanipulation and nanotechnology applied to direct experimental observation of molecular clusters.

- **Numerical Facts.** Numerical simulations are a very useful approach to examine our understanding of the NEAS [146]. Numerical simulations allow one to measure correlation functions and other observables that are hardly accessible in experiments. Susceptibilities in glasses are usually defined in terms of four-point correlation functions (two-point in space and two-point in time), which give information about how spatially separated regions are correlated in time [147]. A characteristic quantity is the typical length of such regions. Numerical simulations of glasses show that the maximum length of spatially correlated regions is small, just a few nanometers in molecular glasses or a few radii in colloidal systems. Its growth in time is also exceedingly slow (logarithmic in time), suggesting that the correlation length is small for the experimentally accessible timescales.
- **Theoretical Facts.** There are several aspects that suggest that glassy dynamics must be understood as a result of the relaxation of CRRs. Important advances in the understanding of glass phenomena come from spin glass theory [148, 149]. Historically, this theory was proposed to study disordered magnetic alloys, which show nonequilibrium phenomena (e.g., aging) below the spin glass transition temperature. However, it has been shown later how spin glass theory provides a consistent picture of the NEAS in structural glass models that do not explicitly contain quenched disorder in the Hamiltonian [150–153]. Most of the progress in this area comes from the study of mean-field models, that is, systems with long range interactions. The success of mean-field theory to reproduce most of the observed phenomenology in glasses suggests that NEASs are determined by the relaxation of mean-field-like regions, perhaps the largest CRRs in the system. Based on this analogy, several mean-field-based phenomenological approaches have been proposed [154–158].

In the next sections I briefly discuss some of the theoretical concepts important to understanding the glass state and nonequilibrium aging dynamics.

A. A Phenomenological Model

To better understand why CRRs are predominantly small, we introduce a simple phenomenological aging model inspired by mean-field theory [155]. The model consists of a set of regions or domains of different sizes s. A region of size s is just a molecular cluster (colloidal cluster), containing s molecules (or s colloidal particles). The system is prepared in an initial high energy configuration, where spatially localized regions in the system contain some stress energy. That energy can be irreversibly released to the environment if a cooperative rearrangement of that region takes place. The release occurs when some correlated structures are built inside the region by a cooperative or anchorage mechanism. Anchorage occurs when all s molecules in that region move to collectively find a transition state that gives access to the *release pathway*, that is, a path in configurational space that activates the rearrangement process. Because the cooperative process involves s particles, the characteristic time to anchor the transition state is given by

$$\frac{\tau_s}{\tau_0} \propto \left(\frac{\tau^*}{\tau_0}\right)^s = \exp\left(\frac{Bs}{T}\right) \tag{196}$$

where $\tau^* = \tau_0 \exp(B/T)$ is the activated time required to anchor one molecule, τ_0 is a microscopic time, and B is the activation barrier that is equal to the energy of the transition state. How do CRRs exchange energy with the environment? Once relaxation starts, regions of all sizes contain some amount of stress energy ready to be released to the environment in the form of heat. The first time a given region rearranges it typically releases an amount of heat $\overline{Q}$ that does not scale with the size of the region. After the first rearrangement has taken place, the region immediately equilibrates with its environment. Subsequent rearrangement events in that same region do not release more stress energy to the environment. These regions can either absorb or release heat from/to the environment as if they were thermally equilibrated with the bath, the net average heat exchanged with the environment being equal to 0. The release of the stored stress energy in the system proceeds in a hierarchical fashion. At a given age t (the time elapsed after relaxation starts, also called waiting time), only the CRRs of size s^* have some stress energy $\overline{Q}$ available to be released to the environment. Smaller regions with $s < s^*$ already released their stress energy sometime in the past, being now in thermal equilibrium with the environment. Larger regions with $s > s^*$ have not yet had enough time to release their stress energy. Only the CRRs with s in the vicinity of s^* contribute to the overall relaxation of the glass toward the supercooled state. That size s^* depends on the waiting time or time elapsed since the relaxation started.

Let $n_s(t)$ be the number of CRRs of size s at time t. At a given time the system is made up of nonoverlapping regions in the system that randomly rearrange according to Eq. (196). After a rearrangement occurs, CRRs destabilize, probably breaking up into smaller regions. In the simplest description we can assume that regions can just gain or lose one particle from the environment with respective (gain,loose) rates k_s^g, k_s^l with $k_s^g + k_s^l = k_s$. k_s, the rate of rearrangement, is proportional to $1/\tau_s$, where τ_s is given in Eq. (196). To further simplify the description, we just take $k_s^g = gk_s, k_s^l = lk_s$ with $g + l = 1$. Consequently, the balance equations involve the following steps:

$$\mathcal{D}_s \rightarrow \mathcal{D}_{s-1} + p; \quad \mathcal{D}_s + p \rightarrow \mathcal{D}_{s+1} \tag{197}$$

with rates k_s^l, k_s^g, where $\mathcal{D}_s$ denotes a region of size s and p denotes a particle (an individual molecule or a colloidal particle) in the system. The balance equations for the occupation probabilities read ($s \geq 2$),

$$\frac{\partial n_s(t)}{\partial t} = k_{s+1}^l n_{s+1}(t) + k_{s-1}^g n_{s-1}(t) - k_s n_s(t) \tag{198}$$

This set of equations must be solved together with mass conservation $\sum_{s=1}^{\infty} s n_s(t) =$ constant. The equations can be solved numerically for all parameters of the model. Particularly interesting results are found for $g \ll l$. Physically, this means that, after rearranging, regions are more prone to lose molecules than to capture them, a reasonable assumption if a cooperative rearrangement leads to a destabilization of the region. A few remarkable results can be inferred from this simple model.

- **Time Dependent Correlation Length.** In Fig. 16a we show the time evolution for $n_s(t)$. At any time it displays a well defined time-dependent cutoff value $s^*(t)$ above which $n_s(t)$ abruptly drops to zero. The distribution of the sizes of the CRRs scales like $n_s(t) = (1/s^*)\hat{n}(s/s^*)$, where s^* is a waiting-time-dependent cutoff size (data not shown). The NEAS can be parameterized by either the waiting time or the size of the region $s^*(t)$. Relaxation to equilibrium is driven by the growth of $s^*(t)$ and its eventual convergence to the stationary solution of Eq. (198). The size $s^*(t)$ defines a characteristic growing correlation length, $\xi(t) = (s^*(t))^{1/d}$, where d is the dimensionality of the system. Because $s^*(t)$ grows logarithmically in time (Eq. (196)), sizes as small as $\simeq 10$ already require 10^{33} iteration steps. Small CRRs govern the relaxation of the system even for exceedingly long times.
- **Logarithmic Energy Decay.** The release of stress energy to the environment occurs when the regions of size s^* rearrange for the first time. The advance of the *front* in $n_s(t)$ located at $s = s^*$ is the leading source of

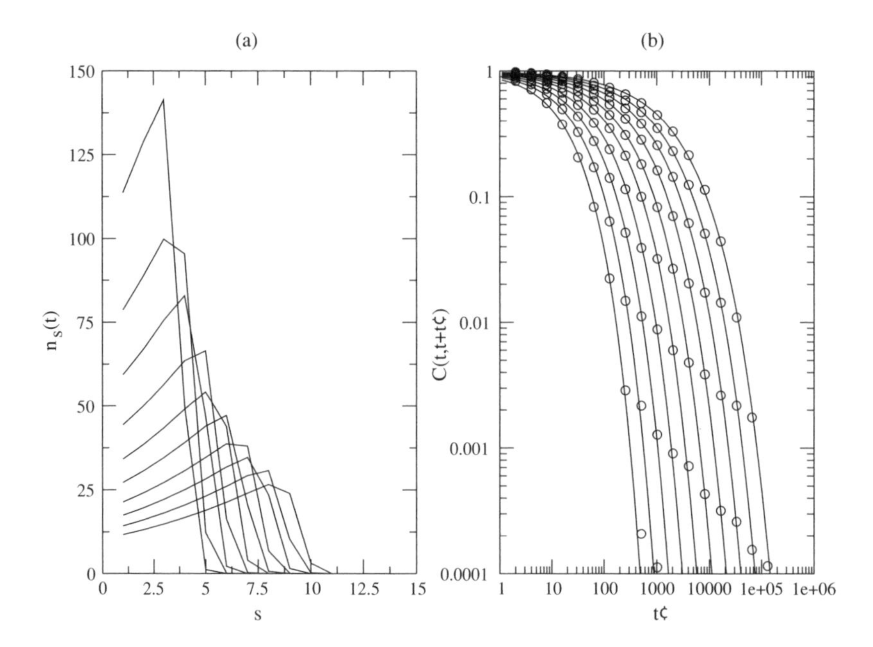

Figure 16. $n_s(t)$ (a) and $C(t, t+t')$ (b) for different waiting times $t = 10^{14}$–10^{33} for the numerical solution of Eq. (198) with $l = 8, g = 1$, and $T = 0.45$. The relaxation time and the stretching exponent are very well fitted by $\tau(t) = 2.2t^{0.35}$, $\beta_s(t) = 0.34 + 0.45t^{-0.06}$. (From Ref. 155.)

energy dissipation. Cooperative rearrangements of regions of size smaller than s^* have already occurred several times in the past and do not yield a net thermal heat flow to the bath, whereas regions of size larger than s^* have not yet released their stress energy. The supercooled state is reached when the cutoff size s^* saturates to the stationary solution of Eq. (198) and the net energy flow between the glass and the bath vanishes. The rate of energy decay in the system is given by the stress energy $\overline{Q}$ released by regions of size $s^*(t)$ times their number $n_{s^*}(t) = \hat{n}(1)/s^*$, divided by the activated time (Eq. (196)) (equal to the waiting time $t \approx \exp(Bs^*/T)$),

$$\frac{\partial E}{\partial t} \approx \frac{\overline{Q} n_{s^*}(t)}{t} \approx \frac{\overline{Q}}{s^* t} \tag{199}$$

Because $s^*(t) \approx T \log(t)$, the energy decays logarithmically with time, $E(t) \approx 1/\log(t)$.

- **Aging.** If we assume independent exponential relaxations for the CRRs, we obtain the following expression for the two-times correlation function:

$$C(t, t+t') = \sum_{s \geq 1} s n_s(t) \exp(-t'/\tau_s) \tag{200}$$

where t denotes the waiting time after the initiation of the relaxation and τ_s is given by Eq. (196). In Fig. 16b we show the correlation function, Eq. (200), for different values of t (empty circles in the figure). Correlations in Eq. (200) are excellently fitted by a stretched exponential with a t-dependent stretching exponent β_s:

$$C(t, t+t') \equiv C_t(t') = \exp\left(-\left(\frac{t'}{\tau_t}\right)^{\beta_s(t)}\right) \tag{201}$$

In Fig. 16b we also show the best fits (continuous lines). Correlation functions show simple aging and scale like t'/t with $t = \exp(s^*/T)$, where s^* is the waiting-time-dependent cutoff size.

- **Configurational Entropy and Effective Temperature.** An important concept in the glass literature that goes back to Adam and Gibbs in the 1950s [159, 160] is the configurational entropy, also called complexity and denoted by $\mathcal{S}_c$ [146]. It is proportional to the logarithm of the number of cooperative regions with a given free energy F, $\Omega(F)$:

$$\mathcal{S}_c(F) = \log(\Omega(F)) \tag{202}$$

At a given time t after relaxation starts, the regions of size s^* contain a characteristic free energy F^*. Fluctuations in these regions lead to rearrangements that release a net amount of heat to the environment, Eq. (199). Local detailed balance implies that, after a rearrangement takes place, new regions with free energies around F^* are generated with identical probability. Therefore,

$$\frac{\mathcal{W}(F \to F')}{\mathcal{W}(F' \to F)} = \frac{\Omega(F')}{\Omega(F)} = \exp(\mathcal{S}_c(F') - \mathcal{S}_c(F)) \tag{203}$$

where $\mathcal{W}(F \to F')$ is the rate of creating a region of free energy F' after rearranging a region of free energy F. Note the similarity between Eqs. (203) and (8). If $\Delta F' = F' - F$ is much smaller than $\mathcal{S}_c(F)$, we can expand the difference in the configurational entropy in Eq. (203) and write

$$\frac{\mathcal{W}(\Delta F)}{\mathcal{W}(-\Delta F)} = \exp\left(\left(\frac{\partial \mathcal{S}_c(F)}{\partial F}\right)_{F=F^*} \Delta F\right) = \exp\left(\frac{\Delta F}{T_{\text{eff}}(F^*)}\right) \tag{204}$$

with the shorthand notation $\mathcal{W}(\Delta F) = \mathcal{W}(F \to F')$ and the time-dependent effective temperature $T_{\text{eff}}(F^*)$ defined as

$$\frac{1}{T_{\text{eff}}(F^*)} = \left(\frac{\partial \mathcal{S}_c(F)}{\partial F}\right)_{F=F^*} \tag{205}$$

In the present phenomenological model, only regions that have not yet equilibrated (i.e., of size $s \geq s^*(t)$) can release stress energy in the form of a net amount of heat to the surroundings. This means that only transitions with $\Delta F < 0$ contribute to the overall relaxation toward equilibrium. Therefore, the rate of energy dissipated by the system can be written as

$$\frac{\partial E}{\partial t} \propto \frac{1}{t} \frac{\int_{-\infty}^{0} dx\, x \mathcal{W}(x)}{\int_{-\infty}^{0} dx\, \mathcal{W}(x)} = \frac{2T_{\text{eff}}(F^*)}{t} \tag{206}$$

where we take

$$\mathcal{W}(\Delta F) \propto \exp\left(\frac{\Delta F}{2T_{\text{eff}}(F^*)}\right) \tag{207}$$

as the solution of Eq. (204). Identifying Eqs. (206) and (199), we get

$$T_{\text{eff}}(F^*) = \frac{2\overline{Q}}{s^*} \tag{208}$$

The time dependence of s^* derived in Eq. (199) shows that the effective temperature decreases logarithmically in time.

B. Nonequilibrium Temperatures

The concept of a nonequilibrium temperature has stimulated a lot of research in the area of glasses. This line of research has been promoted by Cugliandolo and Kurchan in the study of mean-field models of spin glasses [161, 162] that show violations of the fluctuation-dissipation theorem (FDT) in the NEAS. The main result in the theory is that two-time correlations $C(t, t_w)$ and responses $R(t, t_w)$ satisfy a modified version of the FDT. It is customary to introduce the effective temperature through the fluctuation-dissipation ratio (FDR) [163] defined as

$$T_{\text{eff}}(t_w) = \lim_{t_w \to \infty} \left(\frac{\partial C(t, t_w)/\partial t_w}{R(t, t_w)}\right) \tag{209}$$

in the limit where $t - t_w \gg t_w$. In contrast, in the limit $t - t_w \ll t_w$ local equilibrium holds and $T_{\text{eff}}(t_w) = T$. In general, $T_{\text{eff}}(t_w) \geq T$, although there are exceptions to this rule and even negative effective temperatures have been found [164]. These predictions have been tested in many exactly solvable models and numerical simulations of glass formers [146]. In what follows we try to emphasize how the concept of the effective temperature $T_{\text{eff}}(t_w)$ contributes to our understanding of nonequilibrium fluctuations in small systems.

Particularly illuminating in this direction is the study of mean-field spin glasses. These models can be analytically solved in the large volume limit. At the same time, numerical simulations allow one to investigate finite-size effects in detail. Theoretical calculations in mean-field spin glasses are usually carried out by first taking the infinite-size limit and later the long-time limit. Due to the infinite range nature of the interactions, this order of limits introduces pathologies in the dynamical solutions and excludes a large spectrum of fluctuations that are relevant in real systems. The infinite-size limit in mean-field models, albeit physically dubious, is mathematically convenient. Because analytical computations for finite-size systems are not available, we can resort to numerical simulations in order to understand the role of finite-size effects in the NEAS. A spin glass model that has been extensively studied is the random orthogonal model (ROM) [165], a variant of the Sherrington–Kirkpatrick model [166], known to reproduce the ideal mode coupling theory [167]. The model is defined in terms of the following energy function:

$$\mathcal{H} = -\sum_{(i,j)} J_{ij}\sigma_i\sigma_j \tag{210}$$

where the σ_i are N Ising spin variables ($\sigma = \pm 1$) and J_{ij} is a random $N \times N$ symmetric orthogonal matrix with zero diagonal elements. In the limit $N \to \infty$, this model has the same thermodynamic properties as the random-energy model of Derrida [168, 169] or the p-spin model [170] in the large p limit [171, 172]. The ROM shows a dynamical transition at a characteristic temperature $T_{\rm dyn}$ (that corresponds to the mode coupling temperature $T_{\rm MCT}$ in mode coupling theories for the glass transition [173]). Below that temperature, ergodicity is broken and the phase space splits up into disconnected regions that are separated by infinitely high energy barriers. For finite N, the dynamics is different and the dynamical transition is smeared out. The scenario is then much reminiscent of the phenomenological model we discussed in Section VI.A. Different sets of spins collectively relax in finite time scales, each one representing a CRR. There are two important and useful concepts in this regard.

- **The Free Energy Landscape.** An interesting approach to identify CRRs in glassy systems is the study of the topological properties of the potential energy landscape [174]. The slow dynamics observed in glassy systems in the NEAS is attributed to the presence of minima, maxima, and saddles in the potential energy surface. Pathways connecting minima are often separated by large energy barriers that slow down the relaxation. Stillinger and Weber have proposed identifying phase space regions with the so-called inherent structure (IS) [175, 176]. The inherent structure of a region in phase space is the configuration that can be reached by energy minimization starting from any configuration contained in the region.

Inherent structures are used as labels for regions in phase space. Figure 17 (left panel) shows a schematic representation of this concept. Figure 17 (right panel) shows the relaxation of the energy of the inherent structure energy starting from a high energy initial nonequilibrium state [177–179]. Inherent structures are a useful way to keep track of all cooperative rearrangements that occur during the aging process [180].

- **FD Plots.** Numerical tests of the validity of the FDR (Eq. (209)) use fluctuation-dissipation plots (FD plots) to represent the integrated response as a function of the correlation. The integrated version of relation (209) is expressed in terms of the susceptibility,

$$\chi(t, t_w) = \int_{t_w}^{t} dt' \, R(t, t') \tag{211}$$

By introducing Eq. (211) into Eq. (209), we obtain

$$\chi(t, t_w) = \int_{t_w}^{t} dt' \frac{1}{T_{\text{eff}}(t')} \frac{\partial C(t, t')}{\partial t'} = \frac{1}{T_{\text{eff}}(t_w)} (C(t, t) - C(t, t_w)) \tag{212}$$

where we have approximated $T_{\text{eff}}(t')$ by $T_{\text{eff}}(t_w)$. By measuring the susceptibility and the correlation function for a fixed value of t_w and plotting one with respect to the other, the slope of the curve χ with respect to C gives the effective temperature. This result follows naturally from Eq. (212) if we take $C(t, t)$ time independent (which is the case for spin systems). If not, proper normalization of the susceptibility and correlations by $C(t, t)$ is required and a similar result is obtained [181]. A numerical test of these relations in the ROM is shown in Fig. 18. We stress that these results have been obtained in finite-size systems. As the system becomes larger, the time scales required to see rearrangement events become prohibitively longer and the relaxation of the system toward equilibrium drastically slows down.

C. Intermittency

Indirect evidence of nonequilibrium fluctuations due to CRRs in structural glasses has been obtained in Nyquist noise experiments by Ciliberto and co-workers. In these experiments a polycarbonate glass is placed inside the plates of a condenser and quenched at temperatures below the glass transition temperature. Voltage fluctuations are then recorded as a function of time during the relaxation process and the effective temperature is measured:

$$T_{\text{eff}}(\omega, t_w) = \frac{S_Z(\omega, t_w)}{4\mathcal{R}(Z(\omega, t_w))} \tag{213}$$

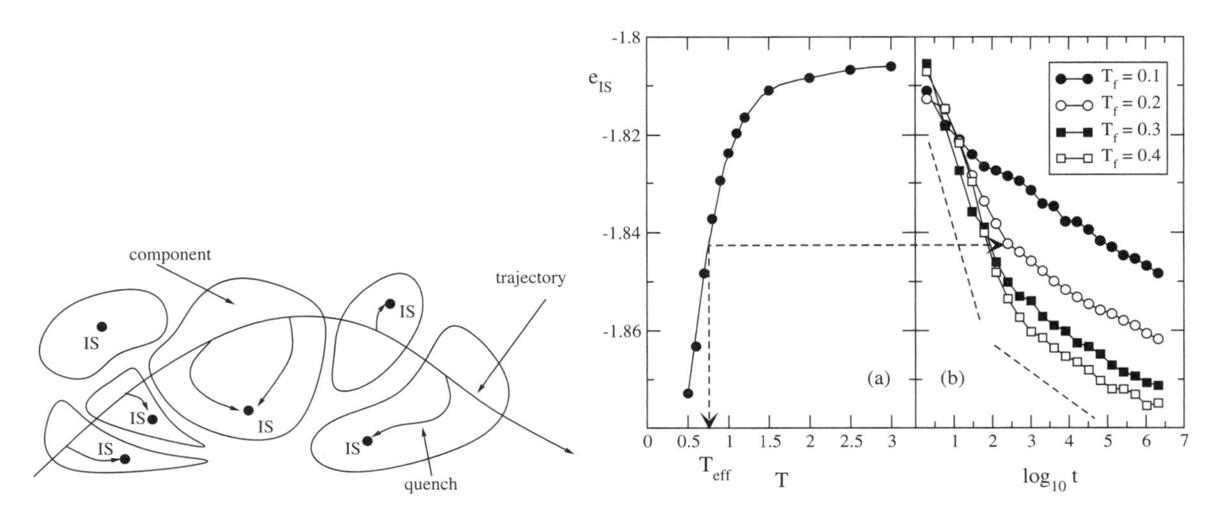

Figure 17. (Left) Stillinger and Weber decomposition. Schematic picture showing regions or components in phase space that are labeled by a given IS. (Right) Relaxation in the ROM. Panel (a): Equilibrium average e_{IS} as a function of temperature. The arrows indicate the construction of the effective temperature $T_{eff}(t_w)$ (Eq. (209)). Panel (b): Average inherent structure energy as function of time for the initial equilibrium temperature $T_i = 3.0$ and final quench temperatures $T_f = 0.1, 0.2, 0.3$, and 0.4. The average is over 300 initial configurations. The system size is $N = 300$. (Left figure from Ref. 180; right figure from Refs. 178 and 179.)

where $\mathcal{R}(Z(\omega, t_w))$ is the real part of the impedance of the system and $S_Z(\omega, t_w)$ is the noise spectrum of the impedance that can be measured from the voltage noise [144].

Experimental data shows a strong variation of the effective temperature with the waiting time by several orders of magnitude. The voltage signal is also intermittent with strong voltage spikes at random times. The distribution of the

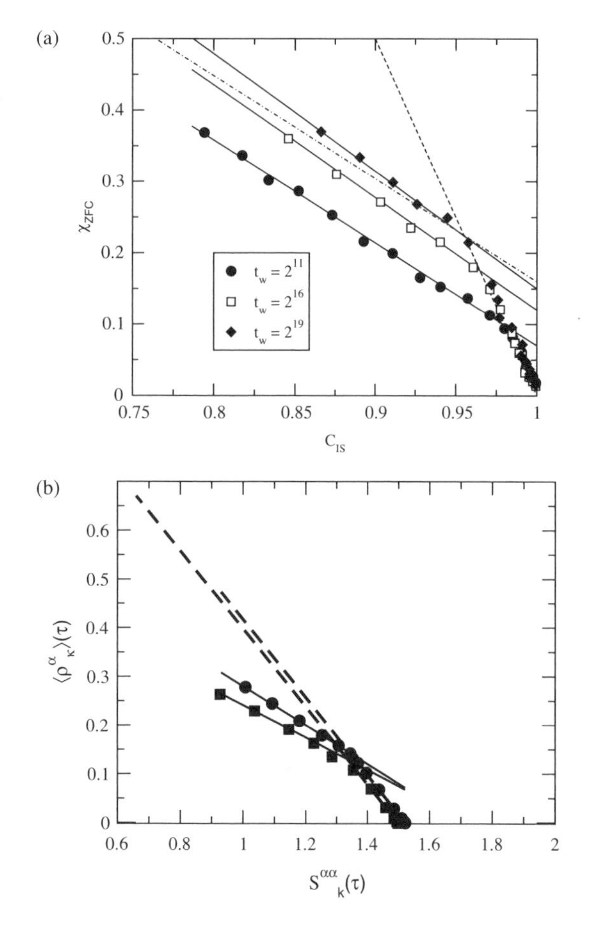

Figure 18. (a) Response versus the dynamical structure factor for the binary mixture Lennard-Jones particles system in a quench from the initial temperature $T_i = 0.8$ to a final temperature $T_f = 0.25$ and two waiting times $t_w = 1024$ (square) and $t_w = 16384$ (circle). Dashed lines have slope $1/T_f$ while thick lines have slope $1/T_{\mathrm{eff}}(t_w)$. (From Ref. 182.) (b) Integrated response function as a function of IS correlation, that is the correlation between different IS configurations for the ROM. The dashed line has slope $T_f = 5.0$, where T_f is the final quench temperature, whereas the full lines are the prediction from Eq. (205) and $F^* = F_{\mathrm{IS}}(T_w) : T_{\mathrm{eff}}(2^{11}) \simeq 0.694, T_{\mathrm{eff}}(2^{16}) \simeq 0.634$, and $T_{\mathrm{eff}}(2^{19}) \simeq 0.608$. The dot-dash line is $T_{\mathrm{eff}}(t_w)$ for $t_w = 2^{11}$ drawn for comparison. (From Ref. 178.)

times between spikes follows a power law characteristic of trap models. These results point to the fact that the observed voltage spikes correspond to CRRs occurring in the polycarbonate sample. Finally, the probability distribution function (PDF) of the voltage signal strongly depends on the cooling rate in the glass, suggesting that relaxational pathways in glasses are very sensitive to temperature changes. A related effect that goes under the name of the Kovacs effect has been also observed in calorimetry experiments, numerical simulations, and exactly solvable models [183–185].

A physical interpretation of the intermittency found in aging systems has been put forward based on exactly solvable models of glasses [186–188]. According to this, energy relaxation in glassy systems follows two different mechanisms (see Section V.C.1): stimulated relaxation and spontaneous relaxation. In the NEAS, the system does not do work but exchanges heat with the environment. Contrary to what was done in previous sections, here we adopt the following convention: $Q > 0$ ($Q < 0$) denotes heat absorbed (released) by the system from (to) the environment. In the NEAS, $\Delta E = Q$: the energy released by the system is dissipated in the form of heat. In the phenomenological model put forward in Section VI.A, different CRRs can exchange (absorb or release) heat to the environment. The regions that cooperatively rearrange for the first time release stress energy to the environment and contribute to the net energy dissipation of the glass. We call this mechanism *spontaneous relaxation*. Regions that have already rearranged for the first time can absorb or release energy from/to the bath several times but do not contribute to the net heat exchanged between the system and the bath. We call this mechanism *stimulated relaxation*. There are several aspects worth mentioning.

- **Heat Distribution.** The distribution of heat exchanges $Q = E(t_w) - E(t)$ for the stimulated process is a Gaussian distribution with zero mean and finite variance. This process corresponds to the heat exchange distribution of the system in equilibrium at the quenching temperature. In contrast, in the spontaneous process a net amount of heat is released to the bath. *Spontaneous heat* arises from the fact that the system has been prepared in a nonequilibrium high energy state. Let us consider a glass that has been quenched at temperature T for an age t_w. During aging, CRRs that release stress energy (in the form of heat $Q < 0$) to the environment satisfy the relation (204):

$$\frac{P^{\rm sp}(Q)}{P^{\rm sp}(-Q)} = \exp\left(\frac{Q}{T_{\rm eff}(F^*)}\right) \tag{214}$$

Therefore, as in the phenomenological model (Eq. (207)), we expect

$$P^{\rm sp}(Q) \propto \exp\left(\frac{Q}{2T_{\rm eff}(F^*)}\right) \quad \text{for} \quad Q < 0 \tag{215}$$

Note that $T_{\text{eff}}(F^*)$ depends on the age of the system through the value of the typical free energy of the CRRs that release their stress energy at t_w, $F^*(t_w)$. This relation has been tested numerically in the ROM (Eq. (210)) by carrying out aging simulations at different temperatures and small sizes N [186] (see next item).

- **Numerical Tests.** How do we measure the heat distribution (Eq. (215)) in numerical simulations of NEAS? A powerful procedure that uses the concept of inherent structures goes as follows. The heat exchanged during the time interval $[t_w, t]$ $(t > t_w)$ has to be averaged over many aging paths (ideally an infinite number of paths). Along each aging path many rearrangement events occur between t_w and t. Most of them are stimulated, a few of them are spontaneous. In fact, because the spontaneous process gets contributions only from those cooperative regions that rearrange for the first time, its PDF signal gets masked by the much larger one coming from the stimulated component where rearrangement events from a single region contribute more than once. To better disentangle both processes, we measure, for a given aging path, the heat exchange corresponding to the first rearrangement event observed after t_w. To identify a rearrangement event, we keep track of the IS corresponding to the run time configuration. Following the IS is an indirect way of catching rearranging events due to CRRs. Only when the system changes IS do we know that a cooperative rearrangement event has taken place. Rearrangement events take place at different times t after t_w, therefore, the heat distribution $P^{\text{sp}}(Q)$ is measured along a heterogeneous set of time intervals. The results for the heat distributions at various ages t_w are shown in Fig. 19. We notice the presence of two well defined contributions to the heat PDFs: a Gaussian central component plus additional exponential tails at large and negative values of Q. The Gaussian component corresponds to the stimulated process; however, its mean is different from zero. The reason for this apparent discrepancy lies in the numerical procedure used to measure the heat PDF: the average *stimulated heat* is not equal to the net exchanged heat (which should be equal to 0) because different aging paths contribute to the heat exchange along different time intervals. The Gaussian component should be equal to the heat PDF for the system in thermal equilibrium at the same temperature and therefore independent of t_w. Indeed, the variance of the Gaussian distribution is found to be independent of t_w [186].
- **Spontaneous Events Release Stress Energy.** One striking aspect of the spontaneous process is that, according to Eq. (214), the probability of heat absorption $(Q > 0)$ should be much larger than the probability of heat release $(Q < 0)$. However, this is not observed in the numerical results of

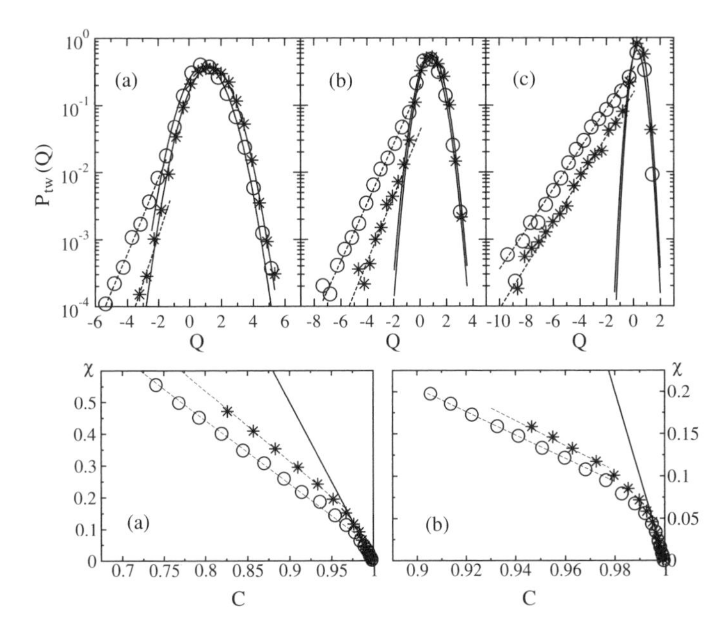

Figure 19. Heat exchange PDFs for $T = 0.3$ (a), $T = 0.2$ (b), and $T = 0.1$ (c). Circles are for $t_w = 2^{10}$ and asterisks for $t_w = 2^{15}$. The continuous lines are Gaussian fits to the stimulated sector; the dashed lines are the exponential fits to the spontaneous sector. (From Ref. 186.)

Fig. 19, where the exponential tail is restricted to the region $Q < 0$. Why are spontaneous events not observed for $Q > 0$? The reason is that spontaneous events can only release and not absorb energy from the environment; see Eq. (215). This is in line with the argumentation put forward in Section VI.A, where the first time that cooperative regions release the stress energy, it gets irreversibly lost as heat in the environment. As the number of stressed regions monotonically decreases as a function of time, the weight of the heat exponential tails decreases with the age of the system as observed in Fig. 19. The idea that only energy decreasing events contribute to the effective temperature (Eq. (215)) makes it possible to define a time-dependent configurational entropy [189].

- **Zero-Temperature Relaxation.** This interpretation rationalizes the aging behavior found in exactly solvable entropy barrier models that relax to the ground state and show aging at zero temperature [190, 191]. At $T = 0$, the stimulated process is suppressed (microscopic reversibility, Eq. (8), does not hold), and Eq. (204) holds by replacing the free energy of a CRR by its energy, $F = E$. In these models a region corresponds to just a configuration in phase space and relaxation occurs through spontaneous rearrangements, where configurations are visited only once. In entropic barrier models the effective temperature (Eq. (205)) still governs aging at $T = 0$. Because the energy is a monotonically decreasing quantity for all aging paths, Eq. (204) does not strictly hold as $\mathcal{W}(\Delta E > 0) = 0$. Yet, the effective temperature obtained from Eq. (204) has been shown to coincide with that derived from the FDR (Eq. (209)) [187, 188].

VII. CONCLUSIONS AND OUTLOOK

We have presented a general overview of several topics related to the nonequilibrium behavior of small systems: from fluctuations in mesoscopic systems such as small beads in optical traps up to molecular machines and biomolecules. The main common theme is that, under appropriate conditions, physical systems exchange small amounts of energy with the environment, leading to large fluctuations and strong deviations from the average behavior. We call such systems *small* because their properties and behavior are markedly different from macroscopic systems. We started our discussion by stressing the similarities between colloidal systems and molecular machines: intermittency and nonequilibrium behavior are common aspects there. We then discussed fluctuation theorems (FTs) in detail and focused our discussion on two well studied systems: the bead in a trap and single molecule force experiments. Experimental results in such systems show the presence of large tails in heat and work distributions in marked contrast to Gaussian distributions, characteristic of macroscopic systems. Such behavior can be rationalized by introducing a path formalism that quantifies work/heat distributions. Finally, we revised some of the main concepts in glassy dynamics where small energy fluctuations appear as an essential underlying ingredient of the observed slow relaxation. Yet, we still lack a clear understanding of the right theory that unifies all phenomena, and a clear and direct observation of the postulated small and cooperatively rearranging regions remains an experimental challenge. We envision three future lines of research.

- **Developments in FTs.** FTs are simple results that provide a new view to better understanding issues related to irreversibility and the second law of thermodynamics. The main assumption of FTs is microscopic reversibility or local equilibrium, an assumption that has received some criticism [192–194]. Establishing limitations on the validity of FTs is the next task for the future. At present, no experimental result contradicts any of the FTs, mainly because the underlying assumptions are respected in the experiments or because current techniques are not accurate enough to detect systematic discrepancies. Under some experimental conditions, we might discover that microscopic reversibility breaks down. We then might need a more refined and fundamental description of the relevant degrees of freedom in the system. Validation of FTs under different and far from equilibrium conditions will be useful to test the main assumptions.

- **Large Deviation Functions.** The presence of large tails can be investigated in statistical mechanics theories by exact analytical solutions of simple models, by introducing simplified theoretical approaches or even by designing smart and efficient algorithms. In all cases, we expect to obtain a good theoretical understanding of the relation between large deviations and nonequilibrium processes. Ultimately, this understanding can be very important in biological systems where nonequilibrium fluctuations and biological function may have gone hand in hand during biological evolution on Earth over the past 4.5 billion years. A very promising line of research in this area will be the study of molecular motors, where the large efficiency observed at the level of a single mechanochemical cycle might be due to a very specific adaptation of the molecular structure of the enzyme to the aqueous environment. This fact may have important implications at the level of single molecules and larger cellular structures.
- **Glassy Systems.** We still need to have direct and clear experimental evidence of the existence of the cooperatively rearranging regions, responsible for most of the observed nonequilibrium relaxational properties in glasses. However, the direct observation of these regions will not be enough. It will also be necessary to have a clear idea of how to identify them in order to extract useful statistical information that can be interpreted in the framework of a predictive theory. Numerical simulations will be very helpful in this regard. If the concept of nonequilibrium temperature has to survive the time then it will be necessary also to provide accurate experimental measurements at the level of what we can now get from numerical simulations.

Since the discovery of Brownian motion in 1827 by the biologist Robert Brown and the later development of the theory for Brownian motion in 1905, science has witnessed an unprecedented convergence of physics toward biology. This was anticipated several decades ago by Erwin Schrödinger, who in his famous 1944 monograph entitled *What Is Life?* [195] wrote when talking about the motion of a clock: "The true physical picture includes the possibility that even a regularly going clock should all at once invert its motion and, working backward, rewind its own spring—at the expense of the heat of the environment. The event is just 'still a little less likely' than a 'Brownian fit' of a clock without driving mechanism." Biological systems seem to have exploited thermal fluctuations to build new molecular designs and structures that efficiently operate out of equilibrium at the molecular and cellular levels [196–199]. The synergy between structure and function is most strong in living systems where nonequilibrium fluctuations are at the root of their amazing and rich behavior.

VIII. LIST OF ABBREVIATIONS

CFT	Crooks fluctuation theorem
CRRs	Cooperatively rearranging regions
JE	Jarzynski equality
FDR	Fluctuation-dssipation ratio
FDT	Fluctuation-dissipation theorem
FEC	Force-extension curve
FT	Fluctuation theorem
IS	Inherent structure
NEAS	Nonequilibrium aging state
NESS	Nonequilibrium steady state
NETS	Nonequilibrium transient state
PDF	Probability distribution function
ROM	Random orthogonal model

Acknowledgments

I am grateful to all my collaborators and colleagues, too numerous to mention, from whom I learned in the past and who have made possible the discussion of the many of the topics covered in this chapter. I acknowledge financial support from the Spanish Ministerio de Eduación y Ciencia (Grant FIS2004-3454 and NAN2004-09348), the Catalan government (Distinció de la Generalitat 2001-2005, Grant SGR05-00688), the European community (STIPCO network), and the SPHINX ESF program.

References

1. T. L. Hill, *Thermodynamics of Small Systems*, Dover Publications, New York, 1994.
2. D. H. E. Gross, *Microcanonical Thermodynamics, Phase Transitions in "Small Systems,"* World Scientific Lecture Notes in Physics, World Scientific, Singapore, 2001.
3. F. Ritort, Work fluctuations, transient violations of the second law and free-energy recovery methods. *Semin. Poincaré* **2**, 193–226 (2003).
4. C. Bustamante, J. Liphardt, and F. Ritort, The nonequilibrium thermodynamics of small systems. *Phys. Today* **58**, 43–48 (2005).
5. E. Frey and K. Kroy, Brownian motion: a paradigm of soft matter and biological physics. *Ann. Phys (Leipzig)* **14**, 20–50 (2005).
6. D. Reguera, J. M. Rubí, and J. M. G. Vilar, The mesoscopic dynamics of thermodynamic systems. *J. Phys. Chem.* **109**, 21502–21515 (2005).
7. H. Qian, Cycle kinetics, steady state thermodynamics and motors—a paradigm for living matter physics. *J. Phys. (Condensed Matter)* **17**, S3783–S3794 (2005).

8. L. Cipelletti and L. Ramos, Slow dynamics in glassy soft matter. *J. Phys. (Condensed Matter)* **17**, R253–R285 (2005).
9. P. N. Pusey and W. Van Megen, Phase behavior of concentrated suspensions of nearly hard colloidal sphres. *Nature* **320**, 340–342 (1986).
10. E. R. Weeks, J. C. Crocker, A. C. Levitt, A. Schofield, and D. A. Weitz, Three-dimensional direct imaging of structural relaxation near the colloidal glass transition. *Science* **287**, 627–631 (2000).
11. R. E. Courtland and E. R. Weeks, Direct visualization of ageing in colloidal glasses. *J. Phys. (Condensed Matter)* **15**, S359–S365 (2003).
12. B. Alberts, D. Bray, A. Johnson, J. Lewis, M. Raff, K. Roberts, and P. Walter, *Essential Cell Biology*. Garland Publishing, New York, 1998.
13. J. D. Watson, T. A. Baker, S. P. Bell, A. Gann, M. Levine and R. Losick, *Molecular Biology of the Gene*. Benjamin Cummings, San Francisco, 2004.
14. E. Eisenberg and T. L. Hill, Muscle contraction and free energy transduction in biological systems. *Science* **277**, 999–1006 (1985).
15. N. J. Córdova, B. Ermentrout, and G. F. Oster, Dynamics of single-motor molecules: the thermal ratchet model, *Proc. Natl. Acad. Sci. USA* **89**, 339–343 (1992).
16. F. Jülicher, A. Adjari, and J. Prost, Modeling molecular motors, *Rev. Modern Phys.* **69**, 1269–1281 (1998).
17. M. E. Fisher and A. B. Kolomeisky, The force exerted by a molecular motor. *Proc. Natl. Acad. Sci. USA* **96**, 6597–6602 (1999).
18. J. Gelles and R. Landick, RNA polymerase as a molecular motor. *Cell* **93**, 13–16 (1998).
19. A. Klug, A marvelous machine for making messages, *Science* **292**, 1844–1846 (2001).
20. G. Orphanides and D. Reinberg, A unified theory of gene expression. *Cell* **108**, 439–451 (2002).
21. M. D. Wang, M. J. Schnitzer, H. Yin, R. Landick, J. Gelles, and S. M. Block, Force and velocity measured for single molecules of RNA polymerase, *Science* **282**, 902–907 (1998).
22. R. J. Davenport, G. J. Wuite, R. Landick, and C. Bustamante, Single-molecule study of transcriptional pausing and arrest by *E. coli* RNA polymerase. *Science* **287**, 2497–2500 (2000).
23. J. W. Shaevitz, E. A. Abbondanzieri, R. Landick, and S. M. Block, Backtracking by single RNA polymerase molecules observed at near-base-pair resolution. *Nature* **426**, 684–687 (2003).
24. E. A. Abbondanzieri, W. J. Greenleaf, J. W. Shaevitz, R. Landick, and S. M. Block, Direct observation of base-pair stepping by RNA polymerase. *Nature* **438**, 460–465 (2005).
25. N. R. Forde, D. Izhaky, G. R. Woodcock, G. J. L. Wuite and C. Bustamante, Using mechanical force to probe the mechanism of pausing and arrest during continuous elongation by *Escherichia coli* RNA polymerase. *Proc. Natl. Acad. Sci. USA* **99**, 11682–11687 (2002).
26. E. G. D. Cohen, D. J. Evans, and G. P. Morriss, Probability of second law violations in shearing steady states. *Phys. Rev. Lett.* **71**, 2401–2404 (1993).
27. D. J. Evans and D. J. Searles, Equilibrium microstates which generate second law violating steady-states. *Phys. Rev. E* **50**, 1645–1648 (1994).
28. G. Gallavotti and E. G. D. Cohen, Dynamical ensembles in nonequilibrium statistical mechanics. *Phys. Rev. Lett.* **74**, 2694–2697 (1995).
29. E. G. D. Cohen, Some recent advances in classical statistical mechanics, in *Dynamics of Dissipation*, Vol. 597 (P. Garbaczewski and R. Olkiewicz, eds.), Springer-Verlag, Berlin, 2002, p. 7.
30. G. Gallavotti, Nonequilibrium thermodynamics? *Preprint*, arXiv:cond-mat/0301172, 2003.

31. C. Jarzynski, Nonequilibrium equality for free-energy differences, *Phys. Rev. Lett.* **78**, 2690–2693 (1997).
32. J. C. Reid, E. M. Sevick, and D. J. Evans, A unified description of two theorems in nonequilibrium statistical mechanics: the fluctuation theorem and the work relation. *Europhys. Lett.* **72**, 726–730 (2005).
33. D. J. Evans, A nonequilibrium free energy theorem for deterministic systems. *Mol. Phys.* **101**, 1551–1554 (2003).
34. D. J. Evans and D. Searles, The fluctuation theorem. *Adv. Phys.* **51**, 1529–1585 (2002).
35. C. Maes, On the origin and use of fluctuation relations on the entropy. *Semin. Poincaré* **2**, 145–191 (2003).
36. C. Jarzynski, What is the microscopic response of a system driven far from equilibrium. In *Dynamics of Dissipation*, Vol. 597 (P. Garbaczewski and R. Olkiewicz, eds.) Springer-Verlag, Berlin, 2002, pp. 63–82.
37. J. Kurchan, Nonequilibrium work relations, *J. Stat. Mech.* (2007) P07005.
38. J. Kurchan, Fluctuation theorem for stochastic dynamics. *J. Phys. A* **31**, 3719–3729 (1998).
39. J. L. Lebowitz and H. Spohn, A Gallavotti–Cohen type symmetry in the large deviation functional for stochastic dynamics. *J. Stat. Phys.* **95**, 333–365 (1999).
40. U. Seifert, Entropy production along a stochastic trajectory and an integral fluctuation theorem. *Phys. Rev. Lett.* **95**, 040602 (2005).
41. T. Yamada and K. Kawasaki, *Prog. Theor. Phys.* **38**, 1031 (1967).
42. G. N. Bochkov and Y. E. Kuzovlev, Non-linear fluctuation relations and stochastic models in nonequilibrium thermodynamics. I. Generalized fluctuation-dissipation theorem. *Physica A* **106**, 443–479 (1981).
43. C. Maes and M. H. van Wieren, Time-symmetric fluctuations in nonequilibrium systems. *Phys. Rev. Lett.* **96**, 240601 (2006).
44. G. Gallavotti, Chaotic hypothesis: Onsager reciprocity and fluctuation dissipation theorem. *J. Stat. Phys.* **84**, 899 (1996).
45. G. E. Crooks, Entropy production fluctuation theorem and the nonequilibrium work relation for free-energy differences. *Phys. Rev. E* **60**, 2721–2726 (1999).
46. G. E. Crooks, Path-ensemble averages in systems driven far from equilibrium. *Phys. Rev. E* **61**, 2361–2366 (2000).
47. J. G. Kirkwood, Statistical mechanics of fluid mixtures. *J. Chem. Phys.* **3**, 300–313 (1935).
48. R. W. Zwanzig, High-temperature equation of state by a perturbation method. I. Nonpolar gases. *J. Chem. Phys.* **22**, 1420–1426 (1954).
49. C. Jarzynski and D. K. Wojcik, Classical and quantum fluctuation theorems for heat exchange. *Phys. Rev. Lett.* **92**, 230602 (2004).
50. Y. Oono and M. Paniconi, Steady state thermodynamics. *Prog. Theor. Phys. Suppl.* **130**, 29–44 (1998).
51. S. Sasa and H. Tasaki, Steady state thermodynamics. *J. Stat. Phys.* **125**, 125–224 (2006).
52. S. Ciliberto and C. Laroche, An experimental test of the Gallavotti–Cohen fluctuation theorem. *J. Phys. IV (France)* **8**, 215–220 (1998).
53. S. Ciliberto, N. Garnier, S. Hernandez, C. Lacpatia, J. F. Pinton, and G. Ruiz-Chavarria, Experimental test of the Gallavotti–Cohen fluctuation theorem in turbulent flows. *Physica A* **340**, 240–250 (2004).

54. K. Feitosa and N. Menon, Fluidized granular medium as an instance of the fluctuation theorem. *Phys. Rev. Lett.* **92**, 164301 (2004).
55. S. Schuler, T. Speck, C. Tierz, J. Wrachtrup, and U. Seifert, Experimental test of the fluctuation theorem for a driven two-level system with time-dependent rates. *Phys. Rev. Lett.* **94**, 180602 (2005).
56. T. Hatano and S. Sasa, Steady-state thermodynamics of langevin systems. *Phys. Rev. Lett.* **86**, 3463–3466 (2001).
57. G. M. Wang, E. M. Sevick, E. Mittag, D. J. Searles, and D. J. Evans, Experimental demonstration of violations of the second law of thermodynamics for small systems and short timescales. *Phys. Rev. Lett.* **89**, 050601 (2002).
58. D. M. Carberry, J. C. Reid, G. M. Wang, E. M. Sevick, D. J. Searles, and D. J. Evans, Fluctuations and irreversibility: an experimental demonstration of a second-law-like theorem using a colloidal particle held in an optical trap. *Phys. Rev. Lett.* **92**, 140601 (2004).
59. J. C. Reid, D. M. Carberry, G. M. Wang, E. M. Sevick, D. J. Searles, and D. J. Evans, Reversibility in nonequilibrium trajectories of an optically trapped particle. *Phys. Rev. E* **70**, 016111 (2004).
60. O. Mazonka and C. Jarzynski, Exactly solvable model illustrating far-from-equilibrium predictions. *Preprint*, arXiv:cond-mat/9912121.
61. R. Van Zon and E. G. D. Cohen, Extension of the fluctuation theorem. *Phys. Rev. Lett.* **91**, 110601 (2003).
62. R. Van Zon and E. G. D. Cohen, Stationary and transient work-fluctuation theorems for a dragged Brownian particle. *Phys. Rev. E* **67**, 046102 (2003).
63. R. Van Zon and E. G. D. Cohen, Extended heat-fluctuation theorems for a system with deterministic and stochastic forces. *Phys. Rev. E* **69**, 056121 (2004).
64. V. Blickle, T. Speck, L. Helden, U. Seifert, and C. Bechinger, Thermodynamics of a colloidal particle in a time-dependent non-harmonic potential. *Phys. Rev. Lett.* **96**, 070603 (2006).
65. R. Van Zon, S. Ciliberto, and E. G. D. Cohen, Power and heat fluctuations for electric circuits. *Phys. Rev. Lett.* **92**, 313601 (2004).
66. J. Zinn-Justin, *Quantum Field Theory and Critical Phenomena.* International Series of Monographs on Physics, No. 92, Clarnedon Press, 1996.
67. M. Baiesi, T. Jacobs, C. Maes, and N. S. Skantzos, Fluctuation symmetries for work and heat. *Phys. Rev. E* **74**, 021111 (2006).
68. N. Garnier and S. Ciliberto, Nonequilibrium fluctuations in a resistor. *Phys. Rev. E* **71**, 060101(R) (2005).
69. E. H. Trepagnier, C. Jarzynski, F. Ritort, G. E. Crooks, C. Bustamante, and J. Liphardt. Experimental test of Hatano and Sasa's nonequilibrium steady state equality. *Proc. Natl. Acad. Sci.* **101**, 15038–15041 (2004).
70. M. J. Lang and S. M. Block, Resource letter: Lbot-1: laser-based optical tweezers. *Am. J. Phys.* **71**, 201–215 (2003).
71. I. Tinoco Jr. and C. Bustamante, How RNA folds. *J. Mol. Biol.* **293**, 271–281 (1999).
72. D. Thirumalai and C. Hyeon, RNA and protein folding: common themes and variations. *Biochemistry* **44**, 4957–4970 (2005).
73. A. V. Finkelstein, Proteins: structural, thermodynamic and kinetic aspects, in *Slow Relaxations and Nonequilibrium Dynamics* (J. L. Barrat and J. Kurchan, eds.) Springer-Verlag, Berlin, 2004, pp. 650–703.

74. F. Ritort, Single molecule experiments in biological physics: methods and applications. *J. Phys. (Condensed Matter)* **18**, R531–R583 (2006).
75. C. Bustamante, J. C. Macosko, and G. J. L. Wuite, Grabbing the cat by the tail: manipulating molecules one by one. *Nat. Rev. Mol. Cell Biol.* **1**, 130–136 (2000).
76. T. Strick, J. F. Allemand, V. Croquette, and D. Bensimon, The manipulation of single biomolecules. *Phys. Today* **54**, 46–51 (2001).
77. T. R. Strick, M. N. Dessinges, G. Charvin, N. H. Dekker, J. K. Allemand, D. Bensimon, and V. Croquette. Stretching of macromolecules and proteins. *Rep. Prog. Phys.* **66**, 1–45 (2003).
78. C. Bustamante, Y. R. Chemla, N. R. Forde, and D. Izhaky, Mechanical processes in biochemistry. *Annu. Rev. Biochem.* **73**, 705–748 (2004).
79. C. R. Calladine and H. R. Drew, *Understanding DNA: The Molecule & How It Works*, Academic Press, London, 1997.
80. S. B. Smith, Y. Cui, and C. Bustamante, An optical-trap force transducer that operates by direct measurement of light momentum. *Methods Enzymol.* **361**, 134–162 (2003).
81. C. Bustamante, S. B. Smith, J. Liphardt, and D. Smith, Single-molecule studies of DNA mechanics. *Curr. Opin. Struct. Biol.* **10**, 279–285 (2000).
82. J. Marko and S. Cocco, The micromechanics of DNA. *Phys. World* **16**, 37–41 (2003).
83. J. Liphardt, B. Onoa, S. B. Smith, I. Tinoco, Jr., and C. Bustamante, Reversible unfolding of single RNA molecules by mechanical force. *Science* **292**, 733–737 (2001).
84. C. Cecconi, E. A. Shank, C. Bustamante, and S. Marqusee, Direct observation of the three-state folding of a single protein molecule. *Science* **309**, 2057–2060 (2005).
85. S. Cocco, J. Marko, and R. Monasson, Theoretical models for single-molecule DNA and RNA experiments: from elasticity to unzipping. *C. R. Phys.* **3**, 569–584 (2002).
86. M. Manosas, D. Collin, and F. Ritort, Force dependent fragility in RNA hairpins. *Phys. Rev. Lett.* **96**, 218301 (2006).
87. D. Keller, D. Swigon, and C. Bustamante, Relating single-molecule measurements to thermodynamics. *Biophys. J.* **84**, 733–738 (2003).
88. M. Manosas and F. Ritort, Thermodynamic and kinetic aspects of RNA pulling experiments. *Biophys. J.* **88**, 3224–3242 (2005).
89. J. D. Wen, M. Manosas, P. T. X. Li, S. B. Smith, C. Bustamante, F. Ritort, and I. Tinoco, Jr., Mechanical unfolding of single RNA hairpins. I. Effect of machine setup on kinetics. *Biophys. J.* **92**, 2996–3009 (2007).
90. M. Manosas, J. D. Wen, P. T. X. Li, S. B. Smith, C. Bustamante, I. Tinoco, Jr., and F. Ritort, Mechanical unfolding of single RNA hairpins. II. Modeling hopping experiments. *Biophys. J.* **92**, 3010–3021 (2007).
91. J. Greenleaf, T. Woodside, A. Abbondanzieri, and S. Block. Passive all-optical clamp for high-resolution laser trapping. *Phys. Rev. Lett.* **95**, 208102 (2005).
92. I. Tinoco, Jr. and C. Bustamante, The effect of force on thermodynamics and kinetics of single molecule reactions. *Biophys. Chem.* **102**, 513–533 (2002).
93. I. Tinoco, Jr., Force as a useful variable in reactions: unfolding RNA. *Ann. Rev. Biophys. Biomol. Struct.* **33**, 363–385 (2004).
94. G. Hummer and A. Szabo, Free-energy reconstruction from nonequilibrium single-molecule experiments. *Proc. Natl. Acad. Sci. USA* **98**, 3658–3661 (2001).
95. C. Jarzynski, How does a system respond when driven away from thermal equilibrium? *Proc. Natl. Acad. Sci. USA* **98**, 3636–3638 (2001).

96. J. M. Schurr and B. S. Fujimoto, Equalities for the nonequilibrium work transferred from an external potential to a molecular system: analysis of single-molecule extension experiments. *J. Phys. Chem. B* **107**, 14007–14019 (2003).
97. F. Douarche, S. Ciliberto, and A. Petrosyan, An experimental test of the Jarzynski equality in a mechanical experiment. *Europhys. Lett.* **70**, 593–598 (2005).
98. F. Douarche, S. Ciliberto, and A. Petrosyan, Estimate of the free energy difference in mechanical systems from work fluctuations: experiments and models. *J. Stat. Mechanics (Theor. Exp.)*, P09011 (2005).
99. U. Gerland, R. Bundschuh, and T. Hwa, Force-induced denaturation of RNA. *Biophys. J.* **81**, 1324–1332 (2001).
100. J. Liphardt, S. Dumont, S. B. Smith, I. Tinoco, Jr., and C. Bustamante, Equilibrium information from nonequilibrium measurements in an experimental test of the Jarzynski equality. *Science* **296**, 1833–1835 (2002).
101. F. Ritort, C. Bustamante, and I. Tinoco, Jr., A two-state kinetic model for the unfolding of single molecules by mechanical force. *Proc. Natl. Acad. Sci. USA* **99**, 13544–13548 (2002).
102. D. Collin, F. Ritort, C. Jarzynski, S. B. Smith, I. Tinoco, Jr., and C. Bustamante, Verification of the crooks fluctuation theorem and recovery of RNA folding free energies. *Nature* **437**, 231–234 (2005).
103. J. Gore, F. Ritort, and C. Bustamante, Bias and error in estimates of equilibrium free-energy differences from nonequilibrium measurements. *Proc. Natl. Acad. Sci. USA* **100**, 12564–12569 (2003).
104. R. H. Wood, W. C. F. Miihlbauer, and P. T. Thompson, Systematic errors in free energy perturbation calculations due to a finite sample of configuration space: sample-size hysteresis. *J. Phys. Chem.* **95**, 6670–6675 (1991).
105. D. M. Zuckermann and T. B. Woolf, Theory of systematic computational error in free energy differences. *Phys. Rev. Lett.* **89**, 180602 (2002).
106. B. Isralewitz, M. Gao, and K. Schulten, Steered molecular dynamics and mechanical functions of proteins. *Curr. Opin. Struct. Biol.* **11**, 224–230 (2001).
107. M. O. Jensen, S. Park, E. Tajkhorshid, and K. Schulten, Energetics of glycerol conduction through aquaglyceroporin glpf. *Proc. Natl. Acad. Sci. USA* **99**, 6731–6736 (2002).
108. S. Park, F. Khalili-Araghi, E. Tajkhorshid, and K. Schulten, Free-energy calculation from steered molecular dynamics simulations using Jarzynski's equality. *J. Phys. Chem. B* **119**, 3559–3566 (2003).
109. I. Andriocioaei, A. R. Dinner, and M. Karplus, Self-guided enhanced sampling methods for thermodynamic averages. *J. Chem. Phys.* **118**, 1074–1084 (2003).
110. S. Park and K. Schulten, Calculating potentials of mean force from steered molecular dynamics simulation. *J. Chem. Phys.* **13**, 5946–5961 (2004).
111. G. Hummer and A. Szabo, Free-energy surfaces from single-molecule force spectroscopy. *Acc. Chem. Res.* **38**, 504–513 (2005).
112. G. Hummer, Fast-growth thermodynamic integration: error and efficiency analysis. *J. Chem. Phys.* **114**, 7330–7337 (2001).
113. D. A. Hendrix and C. Jarzynski, A "fast growth" method of computing free-energy differences. *J. Chem. Phys.* **114**, 5974–5981 (2001).
114. D. M. Zuckermann and T. B. Woolf, Overcoming finite-sampling errors in fast-switching free-energy estimates: extrapolative analysis of a molecular system. *Chem. Phys. Lett.* **351**, 445–453 (2002).

115. O. Braun, A. Hanke, and U. Seifert, Probing molecular free energy landscapes by periodic loading. *Phys. Rev. Lett.* **93**, 158105 (2004).

116. A. Imparato and L. Peliti, Evaluation of free energy landscapes from manipulation experiments. *J. Stat. Mechanics (Theor. Exp.)*, P03005 (2006).

117. F. Ritort, Work and heat fluctuations in two-state systems: a trajectory thermodynamics formalism. *J. Stat. Mechanics (Theor. Exp.)*, P10016 (2004).

118. C. Jarzynski, Rare events and the convergence of exponentially averaged work values. *Phys. Rev. E* **73**, 046105 (2006).

119. I. Kosztin, B. Barz, and L. Janosi, Calculating potentials of mean force and diffusion coefficients from nonequilibrium processes without Jarzynski equality. *J. Chem. Phys.* **124**, 064106 (2006).

120. R. Delgado-Buscaiolini, G. De Fabritiis, and P. V. Coveney, Determination of the chemical potential using energy-biased sampling. *J. Chem. Phys.* **123**, 054105 (2005).

121. C. H. Bennett, Efficient estimation of free-energy differences from Monte Carlo data. *J. Comput. Phys.* **22**, 245–268 (1976).

122. M. R. Shirts, E. Bair, G. Hooker, and V. S. Pande, Equilibrium free energies from nonequilibrium measurements using maximum-likelihood methods. *Phys. Rev. Lett.* **91**, 140601 (2003).

123. P. Maragakis, M. Spithchy, and M. Karplus, Optimal estimates of free energy estimates from multi-state nonequilibrium work data. *Phys. Rev. Lett.* **96**, 100602 (2006).

124. R. C. Lhua and A. Y. Grossberg, Practical applicability of the Jarzynski relation in statistical mechanics: a pedagogical example. *J. Phys. Chem. B* **109**, 6805–6811 (2005).

125. I. Bena, C. Van den Broeck, and R. Kawai, Jarzynski equality for the Jepsen gas. *Europhys. Lett.* **71**, 879–885 (2005).

126. T. Speck and U. Seifert, Dissipated work in driven harmonic diffusive systems: general solution and application to stretching rouse polymers. *Eur. Phys. J. B* **43**, 521–527 (2005).

127. G. E. Crooks and C. Jarzynski, On the work distribution for the adiabatic compression of a dilute classical gas. *Phys. Rev. E* **75**, 021116 (2007).

128. B. Cleuren, C. Van den Broeck, and R. Kawai, Fluctuation and dissipation of work in a Joule experiment. *Phys. Rev. Lett.* **96**, 050601 (2006).

129. P. Hanggi, P. Talkner, and M. Borkovec, Reaction-rate theory: fifty years after Kramers. *Rev. Mod. Phys.* **62**, 251–341 (1990).

130. V. I. Melnikov, The Kramers problem: fifty years of development. *Phys. Rep.* **209**, 1–71 (1991).

131. E. Evans, Probing the relationship between force, lifetime, and chemistry in single molecular bonds. *Ann. Rev. Biophys. Biomol. Struct.* **30**, 105–128 (2001).

132. E. Evans and P. Williams, Dynamic force spectroscopy, in *Physics of Bomolecules and Cells*, Vol. LXXV (H. Flyvbjerg, F. Jülicher, P. Ormos, and F. David, eds.) Springer-Verlag, Berlin, 2002, pp. 145–204.

133. A. Imparato and L. Peliti, Work distribution and path integrals in mean-field systems. *Europhys. Lett.* **70**, 740–746 (2005).

134. A. Imparato and L. Peliti, Work probability distribution in systems driven out of equilibrium. *Phys. Rev. E* **72**, 046114 (2005).

135. B. Derrida, J. L. Lebowitz, and E. R. Speer, Free energy functional for nonequilibrium systems: an exactly solvable case. *Phys. Rev. Lett.* **87**, 150601 (2001).

136. B. Derrida, J. L. Lebowitz, and E. R. Speer, Large deviation of the density profile in the symmetric simple exclusion process. *J. Stat. Phys.* **107**, 599–634 (2002).

137. B. Derrida, J. L. Lebowitz, and E. R. Speer, Exact large deviation functional of a stationary open driven diffusive system: the asymmetric exclusion process. *J. Stat. Phys.* **110**, 775–810 (2003).
138. C. Giardina, J. Kurchan, and L. Peliti, Direct evaluation of large-deviation functions. *Phys. Rev. Lett.* **96**, 120603 (2006).
139. J. Jackle, Models of the glass transition. *Rep. Prog. Phys.* **49**, 171–231 (1986).
140. C. A. Angell, Formation of glasses from liquids and biopolymers. *Science* **267**, 1924–1935 (1995).
141. M. D. Ediger, C. A. Angell, and S. R. Nagel, Supercooled liquids and glasses. *J. Phys. Chem.* **100**, 13200–13212 (1996).
142. Proceedings of the Trieste Workshop unifying concepts in glassy physiscs. *J. Phys. (Condensed Matter)*, 12 (1999).
143. M. D. Ediger, Spatially heterogeneous dynamics in supercooled liquids. *Annu. Rev. Phys. Chem.* **51**, 99–128 (2000).
144. L. Buisson, L. Bellon, and S. Ciliberto, Intermittency in aging. *J. Phys. (Condensed Matter)* **15**, S1163 (2003).
145. L. Berthier, G. Biroli, J. P. Bouchaud, L. Cipelletti, D. El Masri, D. L'Hote, F. Ladieu, and M. Pierno, Direct experimental evidence of a growing length scale accompanying the glass transition. *Science* **310**, 1797–1800 (2005).
146. A. Crisanti and F. Ritort, Violations of the fluctuation dissipation in glassy systems: basic notions and the numerical evidence. *J. Phys. A* **36**, R181–R290 (2003).
147. G. Biroli and J. P. Bouchaud, Diverging length scale and upper critical dimension in the mode coupling theory of the glass transition. *Europhys. Lett.* **67**, 21 (2004).
148. M. Mezard, G. Parisi, and M. A. Virasoro, *Spin-Glass Theory and Beyond*. World Scientific, Singapore, 1987.
149. A. P. Young (ed), *Spin Glasses and Random Fields*. World Scientific, Singapore, 1998.
150. J. P. Bouchaud and M. Mezard, Self induced quenched disorder: a model for the glass transition. *J. Phys. I (France)* **4**, 1109–1114 (1994).
151. E. Marinari, G. Parisi, and F. Ritort, Replica field theory for deterministic models. I. Binary sequences with low autocorrelation. *J. Phys. A* **27**, 7615–7646 (1994).
152. L. F. Cugliandolo, J. Kurchan, G. Parisi, and F. Ritort, Matrix models as solvable glass models. *Phys. Rev. Lett.* **74**, 1012–1015 (1995).
153. S. Franz and J. Hertz, Glassy transition and aging in a model without disorder. *Phys. Rev. Lett.* **74**, 2114–2117 (1995).
154. X. Xia and P. G. Wolynes, Fragilities of liquids predicted from the random first order transition theory of glasses. *Proc. Natl. Acad. Sci. USA* **97**, 2990–2994 (2000).
155. A. Crisanti and F. Ritort, A glass transition scenario based on heterogeneities and entropy barriers. *Philos. Mag. B* **82**, 143–149 (2002).
156. J. P. Garrahan and D. Chandler, Coarse-grained microscopic models of glass formers. *Proc. Natl. Acad. Sci. USA* **100**, 9710–9714 (2003).
157. G. Biroli and J. P. Bouchaud, On the Adam–Gibbs–Kirkpatrick–Thirumalai–Wolynes scenario for the viscosity increase in glasses. *J. Chem. Phys.* **121**, 7347–7354 (2004).
158. A. Garriga and F. Ritort, Mode dependent nonequilibrium temperatures in aging systems. *Phys. Rev. E* **72**, 031505 (2005).
159. J. H. Gibbs and E. A. DiMarzio, Nature of the glass transition and the glassy state. *J. Chem. Phys.* **28**, 373–383 (1958).

160. G. Adam and J. H. Gibbs, On the temperature dependence of cooperative relaxation properties in glass-forming liquids. *J. Chem. Phys.* **43**, 139–146 (1965).
161. L. F. Cugliandolo and J. Kurchan, Analytical solution of the off-equilibrium dynamics of a long-range spin-glass model. *Phys. Rev. Lett.* **71**, 173–176 (1993).
162. L. F. Cugliandolo and J. Kurchan, Weak ergodicity breaking in mean-field spin-glass models. *Philos. Mag. B* **71**, 501–514 (1995).
163. L. F. Cugliandolo, J. Kurchan, and L. Peliti, Energy flow, partial equilibration, and effective temperatures in systems with slow dynamics. *Phys. Rev. E* **55**, 3898–914 (1997).
164. P. Mayer, S. Leonard, L. Berthier, J. P. Garrahan, and P. Sollich, Activated aging dynamics and negative fluctuation-dissipation ratios. *Phys. Rev. Lett.* **96**, 030602 (2006).
165. E. Marinari, G. Parisi, and F. Ritort, Replica field theory for deterministic models. II. A non-random spin glass with glassy behaviour. *J. Phys. A* **27**, 7647–7668 (1994).
166. D. Sherrington and A. Kirkpatrick, Solvable model of a spin-glass. *Phys. Rev. Lett.* **35**, 1792–1796 (1975).
167. W. Gotze and L. Sjogren, Relaxation processes in supercooled liquids. *Rep. Prog. Phys.* **55**, 241–376 (1992).
168. B. Derrida, Random-energy model: limit of a family of disordered models. *Phys. Rev. Lett.* **45**, 79–82 (1980).
169. B. Derrida, Random-energy model: an exactly solvable model of disordered systems. *Phys. Rev. B* **24**, 2613–2626 (1981).
170. E. Gardner, Spin glasses with p-spin interactions. *Nuclear Phys. B*, 747–765 (1985).
171. M. Campellone. Some non-perturbative calculations on spin glasses. *J. Phys. A* **28**, 2149–2158 (1995).
172. E. De Santis, G. Parisi and F. Ritort, On the static and dynamical transition in the mean-field Potts glass. *J. Phys. A* **28**, 3025–3041 (1995).
173. J.-P. Bouchaud, L. F. Cugliandolo, J. Kurchan, and M. Mezard. Mode-coupling approximations, glass theory and disordered systems. *Physica A* **226**, 243–73 (1996).
174. M. Goldstein, Viscous liquids and the glass transition: a potential energy barrier picture. *J. Phys. Chem.* **51**, 3728–3739 (1969).
175. F. H. Stillinger and T. A. Weber, Hidden structure in liquids. *Phys. Rev. A* **25**, 978–989 (1982).
176. F. H. Stillinger, Statistical-mechanics of metastable matter — superheated and stretched liquids. *Phys. Rev. E* **52**, 4685–4690 (1995).
177. A. Crisanti and F. Ritort, Potential energy landscape of finite-size mean-field models for glasses. *Europhys. Lett.* **51**, 147 (2000).
178. A. Crisanti and F. Ritort, Activated processes and inherent structure dynamics of finite-size mean-field models for glasses. *Europhys. Lett.* **52**, 640 (2000).
179. A. Crisanti and F. Ritort, Equilibrium and ageing dynamics of simple models for glasses. *J. Phys. (Condensed Matter)* **12**, 6413–6422 (2000).
180. A. Crisanti and F. Ritort, Inherent structures, configurational entropy and slow glassy dynamics. *J. Phys. (Condensed Matter)* **14**, 1381–1395 (2002).
181. S. M. Fielding and P. Sollich, Observable dependence of fluctuation-dissipation relations and effective temperatures. *Phys. Rev. Lett.* **88**, 050603 (2002).
182. F. Sciortino and P. Tartaglia, Extension of the fluctuation-dissipation theorem to the physical aging of a model glass-forming liquid. *Phys. Rev. Lett.* **86**, 107–110 (2001).

183. S. Mossa and F. Sciortino, Crossover (or Kovacs) effect in anaging molecular liquid. *Phys. Rev. Lett.* **92**, 045504 (2004).
184. E. La Nave, S. Mossa, and F. Sciortino, Potential energy landscape equation of state. *Phys. Rev. Lett.* **88**, 225701 (2002).
185. E. M. Bertin, J.-P. Bouchaud, J.-M. Drouffe, and C. Godreche, The Kovacs effect in model glasses. *J. Phys. A* **36**, 10701–10719 (2003).
186. A. Crisanti and F. Ritort, Intermittency of glassy relaxation and the emergence of a nonequilibrium spontanoeus measure in the aging regime. *Europhys. Lett.* **66**, 253 (2004).
187. F. Ritort, Stimulated and spontaneous relaxation in glassy systems, in *Unifying Concepts in Granular Media and Glassy Systems* (A. Fierro A. Coniglio, and H. Hermann, eds.) Springer-Verlag, New York, 2004.
188. F. Ritort, Spontaneous relaxation in generalized oscillator models for glassy dynamics. *J. Phys. Chem. B* **108**, 6893–6900 (2004).
189. G. Biroli and J. Kurchan, Metastable states in glassy systems. *Phys. Rev. E* **64**, 016101 (2001).
190. F. Ritort, Glassiness in a model without energy barriers. *Phys. Rev. Lett.* **75**, 1190–1193 (1995).
191. L. L. Bonilla, F. G. Padilla, and F. Ritort, Aging in the linear harmonic oscillator. *Physica A* **250**, 315–326 (1998).
192. E. G. D. Cohen and D. Mauzerall, A note on the Jarzynski equality. *J. Stat. Mechanics (Theor. Exp.)* P07006 (2004).
193. C. Jarzynski, Nonequilibrium work theorem for a system strongly coupled to a thermal environment. *J. Stat. Mechanics (Theor. Exp.)* P09005 (2004).
194. R. D. Astumian, The unreasonable effectiveness of equilibrium-like theory for interpreting nonequilibrium experiments. *Am. J. Phys.* **74**, 683 (2006).
195. E. Schrödinger, *What Is Life*? Cambridge University Press, Cambride, UK, 1967.
196. C. De Duve, *Vital Dust*: *The Origin and Evolution of Life on Earth*, Perseus Books Group, 1995.
197. W. R. Loewenstein, *The Touchstone of Life*, Oxford University Press, New York, 1999.
198. F. M. Harold, *The Way of the Cell*, Oxford University Press, New York, 2001.
199. H. Flyvbjerg, F. Jülicher, P. Ormos, and F. David (edts), *Physics of Biomolecules and Cells*, Volume Session LXXV, Springer, Berlin, 2002.

GENERALIZED ENTROPY THEORY OF POLYMER GLASS FORMATION

JACEK DUDOWICZ AND KARL F. FREED

The James Franck Institute and the Department of Chemistry, The University of Chicago, Chicago, IL 60637

JACK F. DOUGLAS

Polymers Division, National Institute of Standards and Technology, Gaithersburg, MD 20899

CONTENTS

Advances in Chemical Physics, Volume 137, edited by Stuart A. Rice

I. INTRODUCTION

Many fluids exhibiting complex molecular structure or interactions solidify by glass formation rather than crystallization (see Fig. 1) Glass-forming liquids are of particular interest in fabricating complex structural materials since the shear viscosity η of these liquids can be *tuned* over an enormous range by moderate temperature changes. (For example, η can alter by as much as 15–20 orders of magnitude [1] over a temperature range of a couple of hundred kelvins; see Fig. 2.) At temperatures in the middle of this broad transition between a low viscosity fluid at high temperatures and a glassy solid at low temperatures, viscous glass-forming liquids can be deformed into intricate shapes that can then be frozen into a solid form upon further cooling. The changes in viscosity occur over a working temperature range that depends sensitively on molecular structure and thermodynamic conditions (e.g., pressure), and this sensitivity of η to temperature changes is crucial for determining the properties of glass-forming liquids and for processing these materials. The concept of "fragility" has been introduced by Angell [2–4] to quantify the strength [5] of the temperature dependence of η, and there are excellent recent reviews that discuss this concept and the general properties of glass-forming liquids [6–9].

Solidification by crystallization, on the other hand, occurs almost *discontinuously* with temperature in many liquids (see Fig. 1), so that

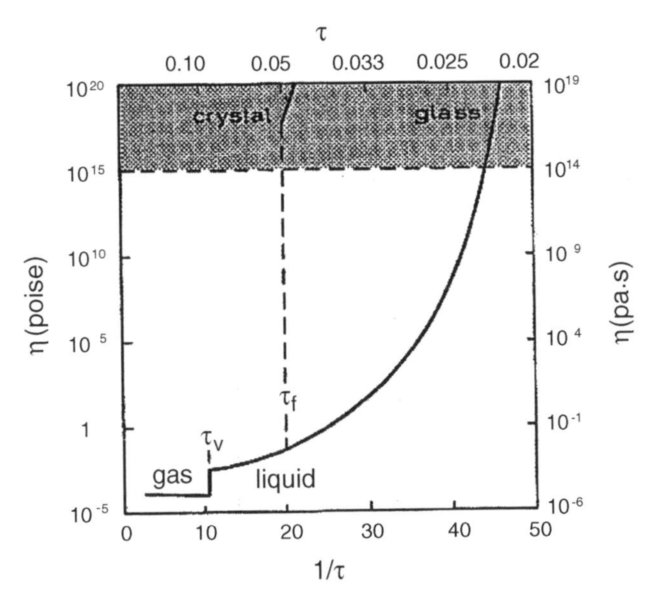

Figure 1. Viscosity η as a function of temperature. The reduced temperature τ is defined as the ratio of thermal energy $k_B T$ and the heat of vaporization, which is a measure of the fluid cohesive energy density. The subscripts on τ indicate the melting point and boiling point, respectively. (Used with permission from Trans Tech Publ. Ltd. W. Kurz and D. J. Fisher, *Fundamentals of Solidification*, 4th ed., Trans Tech Publ Ltd., Switzerland, 1998.)

crystallizing materials must be shaped into form by casting them into molds at very high temperatures, where they are still highly fluid, and then quenching them rapidly to obtain a polycrystalline solid. These highly nonequilibrium amorphous solids strongly reflect the cooling process and casting geometry and, like glass-forming liquids, pose many fundamental problems in their description [1].

Glass formation [2, 3, 6, 7, 10] has been central to fabrication technologies since the dawn of civilization. Glasses not only encompass window panes, the insulation in our homes, the optical fibers supplying our cable TV, and vessels for eating and drinking, but they also include a vast array of "plastic" polymeric materials in our environment [2, 3, 6, 7, 10]. Glasses are also used in high technology applications (e.g., amorphous semiconductors), and there have been recent advances in creating "plastic metallic glasses" potentially suitable for fabricating everyday structural materials [11]. Complex molecular structure and intermolecular interactions are also characteristic features of biological substances, so that many of their properties are likewise governed by the physics of glass formation. This mode of solidification is also prevalent in microemulsions and colloidal suspensions used in numerous commercial products ranging from food additives to cosmetics and shampoo. It is thus no exaggeration to state that glass-forming liquids are ubiquitous in our environment and that their study is one of the must fundamental topics in materials science and biological physics.

Despite the importance of understanding the fundamental nature of glass formation for controlling numerous fabrication processes and material properties and for the preservation of biological substances, a predictive molecular theory for the essential characteristics of glass-forming liquids has been extremely slow to develop. Much of our understanding of the physics of the glass transition derives from experimental studies that have revealed remarkably universal patterns of phenomenology. These striking empirical correlations, in

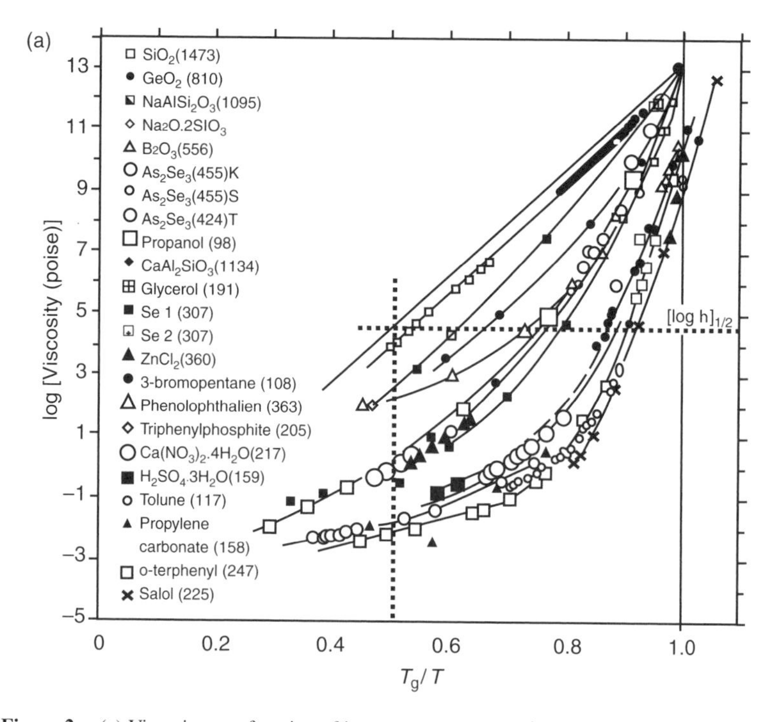

Figure 2. (a) Viscosity as a function of inverse temperature $1/T$ (scaled by the glass transition temperature T_g) for a series of small molecule glass-forming liquids whose relaxation time is specified in parentheses in the key to the figure. The horizontal line in this "fragility" plot is drawn halfway between viscosity η of 10^{12} Pa·s (characteristic for nonfragile liquids) and 10^{-5} Pa·s (the roughly common high temperature value of η). (b) Inverse excess entropy $1/S_{ex}(T)$ (multiplied by the excess entropy $S_{ex}(T_g)$ at the glass transition temperature T_g) as a function of inverse temperature $1/T$ (scaled by T_g) for a series of small molecule glass-forming liquids. (Parts (a) and (b) used with permission from Macmillan Publishers Ltd. L.-M. Martinez and C. A. Angell, *Nature* **410**, 663 (2001). Copyright © 2001.)

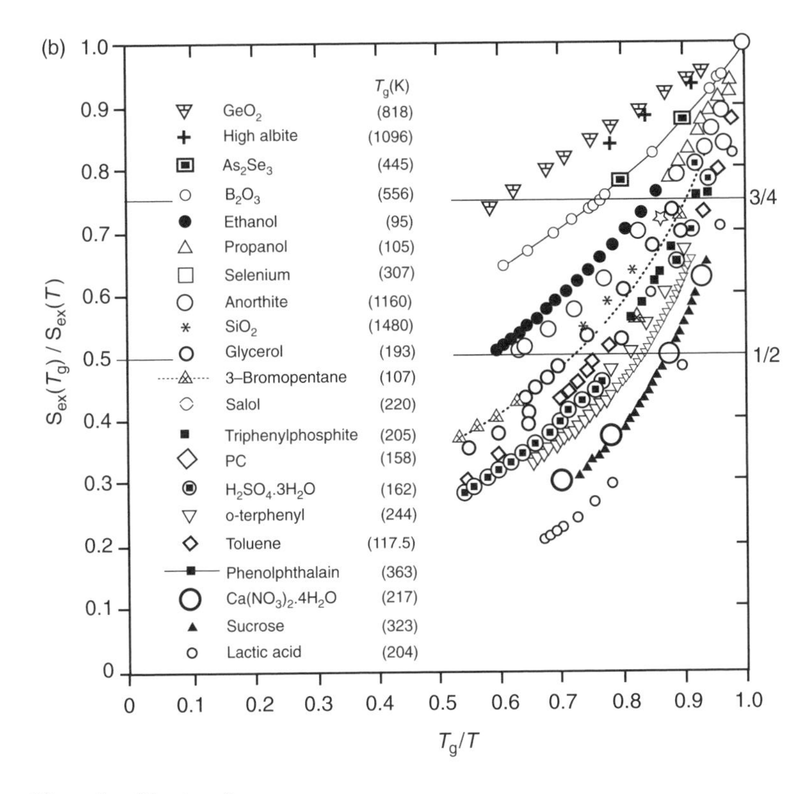

Figure 2. *(Continued)*

conjunction with the practical importance of the topic, naturally invite attempts at a theoretical explanation.

A. Characteristic Temperatures of Glass Formation

Experimental studies have established that long wavelength (large scale) structural relaxation times τ of glass-forming liquids display a nearly Arrhenius temperature dependence at sufficiently high temperatures, but this universal high temperature behavior of liquids breaks down at a temperature T_A well above the glass transition temperature T_g where structural arrest of the fluid occurs. (T_A can exceed 2 T_g, so the deviation from Arrhenius behavior appears far above T_g). The temperature T_A is often observed to be close to the melting temperature if the fluid can crystallize [12–14]. Hence, the temperature range $T < T_A$ is termed the "supercooled" regime, regardless of whether crystallization is observable or not.

The thermodynamic path to glass formation is marked by other characteristic temperatures besides T_A. These characteristic temperatures are universal in glass-forming liquids and serve to define the breadth and nature of glass formation. Below T_A, there is another temperature T_c (or a narrow temperature range) where singular changes in the properties of glass-forming liquids occur. For example, a bifurcation of structural relaxation times τ into fast (β-relaxation) and slow (α-relaxation) modes becomes conspicuous [15] for $T \approx T_c$, and a breakdown in the Stokes–Einstein relation between the diffusion coefficient and the fluid shear viscosity η has been reported [7, 16–18] below this characteristic temperature. Inspired by the mode coupling theory of glass formation, numerous experimental and computational studies estimate T_c by fitting data for τ in the temperature range $T_g \ll T < T_A$ to the relation [19] $\tau \sim (T - T_c)^{-\gamma}$, where T_c and γ are adjustable parameters. Although τ definitely does not diverge for $T = T_c$, the fitted temperature T_c clearly has some physical significance for glass formation since it consistently demarks a narrow temperature range in which numerous properties of the glass-forming fluid alter their character [20]. The importance of T_c is also highlighted by a remarkable reduction of the temperature dependence of τ to a single universal curve that is obtained (see Section IX) by introducing a reduced temperature scale involving T_c and T_g. However, the theoretical foundation for the determination of T_c is rather uncertain. (Mode coupling theory predicts that the "singularity temperature" T_c should lie closer to T_A instead of being roughly midway between T_A and T_g, as found empirically [20, 21].) We identify T_c as the "crossover temperature" T_I between the low and high temperature regimes of glass formation in which fluid transport properties (viscosity, diffusion, relaxation times) display differing temperature dependences [22].

The glass transition temperature T_g is the most commonly studied characteristic of glass-forming liquids, but the physical meaning of this temperature continues to provoke discussion. Operationally, T_g is often determined from the maximum in the specific heat from measurements that are performed at a fixed temperature scanning rate. In reality, these experiments do not provide strictly thermodynamic information, and the observed T_g varies with the temperature scan rate, whether the measurements are performed upon heating or cooling, and other system constraints. Alternatively, T_g is often fixed by the prescription that η at T_g achieves the threshold value of $\eta \approx 10^{14}$ Pa·s or that τ has the order of magnitude $\tau \sim \mathcal{O}(10^2\text{–}10^3 \text{ s})$. Although these criteria are somewhat rough, a narrow temperature range definitely exists near the experimentally determined calorimetric T_g where structural arrest tends to occur. Below T_g glass-forming liquids exhibit a strong tendency to fall out of equilibrium and display noticeable drifts in their properties with time. Thus, nonequilibrium "aging" phenomena become prevalent below T_g, and a thermodynamic description of relaxation becomes inadequate.

The strong temperature dependence of τ below T_c serves to define another characteristic temperature of glass formation, the Vogel temperature T_∞. An astoundingly large class of liquids (ionic, polymeric, biological materials, colloidal dispersions, emulsions, etc.) exhibit an apparent essential singularity in the *extrapolated* temperature dependence of their structural relaxation and other transport properties at low temperatures, that is,

$$\tau \approx \exp[A/(T - T_{\infty})], \quad T_g < T < T_c$$

Of course, this phenomenology [2, 3, 6–10] does *not* imply that τ *actually diverges* as $T \rightarrow T_{\infty}$ any more than the definition of the mode coupling temperature by a power law scaling for $T > T_c$ implies that τ actualy diverges at T_c. Nevertheless, the temperature T_{∞} is subject to a well defined experimental determination and has general significance for characterizing the dynamics of glass-forming liquids [14].

The above discussion reiterates our viewpoint that glass formation involves a relatively broad transition between simple fluid behavior at high temperatures and a nonequilibrium solid at low temperatures. More specifically, the temperature T_A defines the *onset temperature* for this transformation, while the solidification process is basically finished for T_{∞}, where extrapolated relaxation times become astronomical in magnitude. The temperature T_c demarks a transition between distinct high and low temperature regimes of glass formation in which the temperature dependence of τ and other properties are different. The glass transition temperature T_g is clearly some kind of kinetic transition temperature at which the fluid becomes stuck and below which it is very difficult for the system to achieve equilibrium. The theoretical challenge is to calculate these characteristic temperatures and to understand their significance based on a molecular theory.

B. Entropy and Glass Formation

While new theories for the origin and nature of glass formation are constantly being introduced, some ideas remain invariant for qualitatively understanding the nature of this phenomenon. Since the works of Simon [23] and Kauzmann [24], it has generally been appreciated that the rapid increase in viscosity η and structural relaxation times τ associated with glass formation is accompanied by a drop in the fluid entropy S (see Fig. 2b). Indeed, the drops in S upon lowering temperature T are so rapid that extrapolations of S to low temperatures would lead to a negative S at a nonzero temperature (the Kauzmann "paradox"). While a negative entropy is patently absurd for a system in equilibrium, this finding is completely acceptable for nonequilibrium materials, so that the interpretation of this trend in the entropy remains unclear.

Martinez and Angell [4] have recently reviewed the parallelism between the temperature dependence of the viscosity and that of S, and they have shown that this correlation is strong for the small molecule glass formers considered in their study. Since these findings are relevant to the entropy theory of glass formation, we reproduce the reduced temperature dependence of log η and $S(T_g)/S(T)$ in Figs. 2a and 2b, respectively, where S denotes the excess entropy of the fluid relative to that of the crystal. (Angell has popularized the representation of η and other transport data of glass-forming liquids in these types of plot, which accordingly are termed "Angell plots.") The correspondence between log η and $S(T_g)/S(T)$ follows from the Adam–Gibbs theory of glass formation if $S(T)$ is taken to approximate the configurational entropy of the fluid. (Analysis of several small molecule liquids indicates that the vibrational contribution to the excess entropy S_{exc} can be as large as 30% of S_{exc}[25]. We expect these contributions to be even larger for polymer fluids, so that this identification is a rather serious uncontrolled approximation.) Deviations from linearity in Angell plots define the glass fragility in a general sense, and Figs. 2a and 2b illustrate dynamic and thermodynamic fragilities, respectively. Qualitatively, the temperature variations of log η and $S(T_g)/S(T)$ closely track each other, indicating that thermodynamics provides a good qualitative predictor of the relative strength of the temperature dependence of η in these fluids. Unfortunately, more recent work by McKenna and co-workers [26] demonstrates that this correlation between dynamic fragility based on transport measurements and thermodynamic measures of fragility based on $S(T)$ does not hold well for many polymers. We attribute this disturbing result to the inaccuracy of approximating the configurational entropy by the excess entropy of polymer fluids, and this issue is discussed at length later. At any rate, the data in Figs. 2a and 2b point to a striking correlation between η and the excess entropy of small molecule glass formers. It would evidently be useful to develop a generalization of this relation for polymer fluids in a fashion that does not require introducing unverified approximations for the configurational entropy. We next consider some other models of glass formation that are based on functional methods and a molecular description of fluid structure in order to motivate certain aspects of our theory.

C. Integral Equation Approaches to Glass Formation Involving Functionals

Considerations of the dynamics of a fluid composed of simple monatomic particles have led to the development of kinetic models of glass formation and an integral equation description of both the equilibrium structure and dynamic properties of solidifying liquids. To render this approach tractable, all the other particles are assumed to affect the dynamics of a chosen test particle through a mean field effective interaction that describes the molecular crowding that ultimately causes structural arrest. The mode coupling theory of glass formation is the most successful theory [27] of this kind and has simulated numerous experimental and simulation studies since the theory directly addresses experimentally accessible properties and since its theoretical predictions can be subjected to unambiguous tests [28].

Although the derivation of mode coupling theory begins from a formal description for the dynamics of a fluid in terms of projection operators and reasonable molecular interactions, explicit computations invoke *ad hoc* approximations that render the significance of this theory highly uncertain. In practice, the theory focuses on nonlinear integral equations for the reduced dynamic structure factor $\phi(q,t)$ describing the time average of density fluctuations in the fluid at a scattering wavevector $q(\phi(q,t)) = S(q,t)/S(q)$, where $S(q)$ and $S(q,t)$ are the static and dynamic structure factors, respectively. The memory kernel of these integral equations is developed formally in powers of $\phi(q,t)$. Unfortunately, the results depend very strongly on the truncation of this expansion, and no compelling physical arguments justify the uncontrolled truncation approximations adopted. Mode coupling theory for glass formation is entirely distinct from the mode coupling theory for the critical dynamics of phase transitions, which is based on the relation between thermally activated relaxation processes and long wavelength thermodynamic properties ("collective variables"), such as the compressibility [29]. Hence, the latter theory is conceptually closer to our entropy theory of glass formation.

The mode coupling theory of glass formation actually has an antecedent in the Kirkwood–Monroe theory of solidification, which centers on a mean field integral equation for the structure factor $S(q)$ and which identifies solidification with the bifurcation of the solution to this integral equation for those thermodynamic conditions where the fluid becomes unstable [30–33]. The truncation of the expansion for $\phi(q,t)$ in mode coupling theory is the analogue of the "closure" approximation in the integral equation theory and can likewise produce a bifurcation in its solutions for relaxation times (α and β relaxation processes). Mode coupling theory also predicts divergent relaxation times for density fluctuations at a finite characteristic temperature T_c, which is sometimes termed the "ideal glass transition temperature" by practitioners of mode coupling theory. Although it is conceptually interesting that this schematic mode coupling theory predicts the existence of structural arrest at a particular temperature T_c, the theoretically predicted T_c is nowhere near the glass transition temperature T_g, but instead this temperature is found to be closer to T_A [14, 21]. Since T_A often lies near [12] the melting temperature T_m in simple fluids, the latter theory may not be a theory of glass formation at all, but rather may be considered a primitive theory of *crystallization* like the original Kirkwood–Monroe theory.

There is another analytic theory of glass formation that is a descendent of Kirkwood–Monroe theory and that provides some interesting predictions concerning the scattering properties of glass-forming liquids. Klein and coworkers [34–36] have analyzed the pair correlation function for a fluid of cooled particles interacting with an idealized long range, but "soft," potential instead of the hard sphere potential of Kirkwood–Monroe. (The choice of potential ensures the validity of the mean field approximation and is not based on physical considerations.) Klein's model is reported to produce a spinodal instability corresponding to a finite wavevector thermodynamic instability. Specifically, the scattering intensity for this fluid is predicted to diverge at a finite wavevector as in Liebler's mean field type [37] ordering transition in block copolymers [34–36]. Klein and colleagues claim that molecular dynamics simulations for a glass-forming binary Lennard-Jones fluid mixture in two dimensions [35] are consistent with this theory. Interestingly, Klein and co-workers physically identify this spinodal instability with the formation of dynamic clusters ("clumps"), whose long lifetime is suggested to govern the rate of structural relaxation in glass-forming liquids [34]. Simulations for a model fluid with soft long range interactions in three dimensions indeed provide striking evidence for the formation of dynamic clumps at equilibrium. Unfortunately, the extension of MD simulations to a glass-forming fluid in three dimensions with realistic intermolecular interactions [38] has not yet produced convincing agreement with the theory of Klein and colleagues. Nonetheless, the theory of Klein and co-workers seems to be the only molecular theory that directly predicts the onset of the dynamic clustering that is now thought to be an essential aspect of glass-forming liquids, an effect not captured by mode coupling theory. Of course, the mean field theory of Klein and co-workers cannot be expected to faithfully describe the geometrical form of the dynamic clusters that arise in conjunction with glass formation. Section XII summarizes information from experiments and simulations, along with speculation concerning the geometrical form of these dynamic clusters.

An ambitious treatment of glass formation in polymer melts has been developed by Schweizer and Saltzman [39, 40], who basically combine concepts taken from schematic mode coupling theory with the idea of thermal activation processes that are dependent on the compressibility of a polymer melt. Section XI describes some results of this "activation energy" theory and their relation to our entropy theory of glass formation. The predictions of Schweizer–Saltzman theory in the high temperature regime of glass formation seem to have much in common with those predicted by the entropy theory of glass formation.

Finally, the Kirkwood–Monroe theory has also spawned powerful density functional approaches to describe both crystallization [41] and glass formation [42–44]. While highly attractive from a philosophical standpoint, these approaches have so far been limited to treating simple fluids (spherical symmetric particles, such as Lennard-Jones particles or hard spheres) and not complex fluids such as polymers. As with the original Kirkwood–Monroe theory, this class of theories is still preoccupied with the calculation of instability criteria for the liquid state, along with theoretical discussions about the mathematical validity of these conditions. One important concept emerging from the density functional approach to glass formation [42] is the identification of the glass transition temperature T_g with a particle localization–delocalization transition that may be described by a Lindemann relation as for the melting of crystals [42]. Wolynes and co-workers [42, 43] designate this transition as a "random first order transition," presumably due to the similarity between the mathematical mechanisms rather than anything to do with thermodynamics of first order phase transitions. Moreover, they have suggested the existence of "entropic droplets" composed of particles in a relatively delocalized state relative to the background of localized particles. This picture is *quite different* from the integral equation description by Klein and co-workers [34–36], where the "droplets" are relatively *high density* regions of relatively immobile particles. Further arguments [42] that the barrier height for collective particle motion in supercooled liquids can be estimated from the surface energy of these entropic clusters of mobile particles have been discussed critically by Bouchaud and Biroli [45].

D. Generalized Entropy Theory of Glass Formation

In addition to providing a molecular understanding of the universal properties of glass-forming polymer fluids, our recent work seeks to describe the molecular basis for the strong variations in fragility between different polymer materials. Our molecularly based theory is designed to explain the physical meaning of the characteristic temperatures within a theoretically consistent framework and to enable the direct computation of these temperatures in terms of prescribed polymer molecular structures [46]. We also focus on the molecular and thermodynamic factors that control the rate of change of η (or τ) with temperature since this fragility is highly variable for different polymer materials and is one of the most important properties governing applications of glass-forming fluids. The establishment of a physical understanding for the origin of variations in the fragility of glass-forming polymer liquids between materials should enable this property to be manipulated in numerous applications involving glass-forming liquids.

Our recent approach to understanding the fundamental molecular nature of glass formation builds on the foundation of the original entropy theory of glass formation in polymers, developed by Gibbs and DiMarzio [47], and on the Adam–Gibbs relation [48] between the rate of long wavelength structural relaxation τ in glass-forming liquids and the fluid's configurational entropy. The original Gibbs–DiMarzio (GD) approach [47] is a thermodynamic theory and only focuses on estimating an "ideal glass transition temperature" (not to be confused with "ideal glass transition temperature" T_c of mode coupling theory!) where the configurational entropy formally extrapolates to zero within a mean field Flory–Huggins (FH) model of semiflexible polymer chains. Adam and Gibbs [48] (AG), and numerous others following them [2, 49], are largely concerned with estimating τ from thermodynamic information that is obtained from specific heat data. Although GD theory has claimed many successes in rationalizing trends in T_g with molecular structure and AG theory has provided fundamental insights into the rate of change of τ with temperature ("fragility") in small molecule glass-forming liquids [4], surprisingly no previous attempts have been made at *combining* the GD and AG theories.

Instead of estimating the configurational entropy experimentally, as in previous tests of the AG model, we calculate the configurational entropy directly within the lattice cluster theory generalization [50, 51] of the FH mean field approximation for semiflexible polymers. The generalization enables us to treat models of polymers with explicit backbone and side group molecular structures and with variable flexibility and molecular interactions for these chain portions, as well as the dependence on pressure and the addition of solvent [52]. This analytic generalized entropy theory of glass formation emerges from combining our more physically accurate model for the thermodynamics of a polymer fluid with the AG model linking thermodynamics to long wavelength structural relaxation. As shown later, this new entropy theory enables the direct computation of all the characteristic temperatures of glass formation [22, 53], variations of fragility with polymer structure [54], and the explicit computation of τ as a function of molecular parameters, such as monomer structure and differing flexibilities of different portions of the monomers [22].

The analytic generalized entropy theory also considers important issues regarding the meaning of the configurational entropy that should enter into the formal AG expression for τ, issues that have previously been neglected. Past tests of the AG theory are normally predicated [15, 49] on the assumption that the configurational entropy can be *equated* with the excess fluid entropy (obtained from specific heat measurements) by subtracting the entropy of the crystalline or low temperature glass states from the total fluid entropy. This approximation is based on the optimistic presumption that the vibrational contribution to the entropy cancels in this entropy difference. By calculating the configurational entropy theoretically using a lattice model [52], we avoid the uncertainty of this assumed cancellation, which has been the source of much ambiguity in previous attempts to test the AG theory.

A basic question concerns whether the configurational entropy should be defined as an entropy per unit mass (or molar entropy) or as an entropy per unit volume (a site entropy or an "entropy density") [22]. The majority of experimental studies [15, 49] concerning the validity of the AG theory use the molar entropy, while Binder and co-workers [55] have employed the site entropy in their computational studies of the applicability of AG theory. In fact, our calculations indicate [22] that the configurational entropy per unit volume and the configurational entropy per unit mass exhibit quite different temperature dependences and that only the use of the latter in the AG relation for τ leads to a sensible theory of structural relaxation at temperatures much higher than T_g. The distinction between these two normalizations for the configurational entropy is *unimportant* [22], however, in the temperature range between T_g and T_c where much of the relaxation data has been reported.

The development of a theoretical understanding for the dependence of T_g on monomer structure within the generalized entropy theory has also required a deeper reflection concerning how glass formation affects other thermodynamic properties besides the configurational entropy s_c since s_c exhibits no particular signature at T_g. The fluid compressibility, on the other hand, becomes small near T_g, and the fluid particles simple become stuck near this temperature, so that it is difficult for particles to explore available conformational states. This basic problem is resolved in the entropy theory of glass formation by defining T_g in terms of a Lindemann criterion [56] that prescribes a critical value for the mean amplitude of particle displacement relative to the mean interparticle separation in order for fluidity to exist. (This criterion for glass formation has previously been advocated by Xia and Wolynes [42] and by others; see discussion in Section VI.) The microscopic Lindemann "free volume" condition for glass formation naturally translates into a macroscopic free volume constraint that is expressed in terms of the specific volume [53]. Thus, the kinetic transition temperature T_g can be defined [53] by a thermodynamic condition that does not involve the configurational entropy. Our theory has the advantage of providing a description of *all* thermodynamic properties of polymer melts. Consequently, we compare the generalized entropy theory with prior viewpoints of glass formation to test their merits. This flexibility of our approach also allows a comparison of the theory with previous free volume models of glass formation and with the more recent theories of polymer glass formation developed by Schweizer and Saltzman [39, 40], Kivelson and co-workers [13], Dyre et al. [57], and Buchenau and Zorn [58] that all describe the rate of structural relaxation in glass-forming liquids in terms of an activated rate process.

II. STRENGTHS AND WEAKNESSES OF THE ORIGINAL ENTROPY THEORIES OF GLASS FORMATION

A. Strengths and Weaknesses of the Gibbs–DiMarzio Theory

Gibbs and DiMarzio [47] (GD) first developed a systematic statistical mechanical theory of glass formation in polymer fluids, based on experimental observations and on lattice model calculations by Meyer, Flory, Huggins, and others [59]. Apart from providing a clear physical picture of glass formation upon cooling as arising from a vanishing number of accessible configurational states due to the increasing rigidity of polymer chains [47, 60], this theoretical approach has produced a wide array of quantitative predictions regarding glass formation, and numerous successes of the theory are reviewed by DiMarzio and Yang [61, 62]. The main focus of GD theory is the "configurational entropy" S_c, which is that portion of the fluid's entropy relating to the number of distinct configurational states of the fluid alone, with vibrational contributions being excluded. Unfortunately, the conceptual clarity of the notion of configurational entropy S_c is not matched by any direct method for determining this quantity experimentally. Consequently, comparisons of GD theory with experiment often involve uncontrolled approximations associated with attempts to evaluate S_c [63]. On the other hand, recent progress in estimating S_c from numerical simulations has been achieved by determining the number of accessible minima in the potential energy surface describing the glass-forming liquid under constant volume conditions [64–68]. While this numerical estimate of the configurational entropy must be closely related to the S_c calculated from the entropy theory, the exact relation between them has not yet been established. Despite this uncertainty, both of these quantities are conventionally designated by the term configurational entropy.

In addition to the fundamental difficulties of estimating S_c, strong criticism has been raised against fundamental tenets of GD theory. For example, there has been widespread disagreement concerning the identification by GD of a vanishing of S_c with a second order phase transition and of the glass transition temperature with the temperature of this hypothetical transition [69]. Moreover, even the vanishing of S_c has been suggested to be an artifact of the inaccuracy of the mean field calculation of S_c for dense polymer fluids [70]. In the final assessment, however, the *qualitative picture* of polymer glass formation as stemming from an "entropy catastrophe" remains a viable conceptual model [71]. For instance, mean field calculations for spin models of glass formation have converged with results generated from GD theory, suggesting that the entropy catastrophe concept has broad applicability [72].

Simulations [73] have recently provided some insights into the formal $S_c \to 0$ limit predicted by mean field lattice model theories of glass formation. While Monte Carlo estimates of τ for a Flory–Huggins (FH) lattice model of a semiflexible polymer melt extrapolate to infinity near the "ideal" glass transition temperature T_0, where S_c extrapolates to zero, the values of S_c computed from GD theory are too low by roughly a constant compared to the simulation estimates, and this constant shift is suggested to be sufficient to prevent S_c from strictly vanishing [73, 74]. Hence, we can reasonably infer that S approaches a small, but nonzero asymptotic low temperature limit and that S_c similarly becomes *critically small* near T_0. The possibility of a constant residual configurational entropy at low temperatures is briefly mentioned in the original paper by GD and has recently been emphasized again by DiMarzio [62]. Thus, while the *literal* prediction of a vanishing S_c at a finite temperature T_0 is suspect, this formal extension of the thermodynamic theory still retains value as an indicator of an entropy crisis that is identifiable from *extrapolations* of both thermodynamic (e.g., specific heat) and dynamic (e.g., viscosity, diffusion) properties. By considering the configurational entropy of the lattice model to represent the excess configurational entropy relative to the entropy of the glass, the conceptual difficulties of the original GD theory can largely be eliminated.

The original GD theory of glass formation in polymer liquids involves a number of other assumptions and approximations that significantly limit the predictive capacity of the theory. First, the theory is preoccupied with the general philosophical problem of locating and explaining the conceptual basis for the "ideal" glass transition temperature T_0. Unfortunately, a fluid cannot remain in equilibrium near T_0 because the structural relaxation time becomes astronomical in magnitude near this temperature. Hence, comparisons of GD theory with experiment are necessarily indirect. Notably, the experimental glass transition temperature T_g (determined from the maximum in the specific heat or from a change in the slope of the density as a function of temperature [2, 8, 24]) normally occurs [9, 75] above the Kauzmann temperature T_K (where the excess fluid entropy S_{exc} extrapolates to zero), and the difference between T_K and T_g is neglected in the GD theory [76]. Second, the theory is based on a highly simplified description of polymer chains as semiflexible self-avoiding walks composed of structureless monomer units. Thus, little can be said about how monomer and solvent structure affect glass formation. In particular, recent experiments [77–79] have established that more subtle aspects of glass formation, such as fragility [5], depend significantly on the geometry and the degree of rigidity of the polymer side groups. Third, since the complex changes in the dynamics of glass-forming liquids often initiate at temperatures exceeding $2T_g$, it is important to determine the breadth of this transition by estimating the temperatures characterizing the beginning, middle, and end of this broad transition phenomenon. These issues are beyond the predictive capacity of the classic GD theory and are the principal focus of the generalized entropy theory of glass formation.

B. Strengths and Weaknesses of the Adam–Gibbs Theory

The AG model [48] for the dynamics of glass-forming liquids essentially postulates that the drop in S upon lowering temperature is accompanied by collective motion and that the fluid's structural relaxation times τ are activated with a barrier height $\mathcal{E}$ that is proportional [80] to the number z^* of polymer segments within hypothetical "cooperatively rearranging regions" (CRR). (These elements do not necessarily belong to an individual chain, so that similar dynamic structures presumably also arise in low molar mass glass-forming fluids.) At high temperatures, the molecular displacements at an atomic scale are taken to be entirely noncollective, and the AG barrier height $\mathcal{E}_{AG}$ reduces to a (constant) Arrhenius activation energy $\Delta\mu$. More generally, τ is estimated as $\tau = \tau_o \exp(\beta\mathcal{E}_{AG})$, where $\mathcal{E} \approx \mathcal{E}_{AG} \equiv \Delta\mu z^*$ and τ_o is a constant. Extensive experimental estimates of $\Delta\mu$ exist for both small molecule [81] and polymer liquids [82]. For instance, data for the viscosity and tracer diffusion in weakly supercooled alkane liquids indicate a nearly linear dependence of $\Delta\mu$ on polymer mass for a relatively low molar mass range (6–16 carbon atoms) [83, 84]. A proportional increase of the apparent activation energy with polymer mass has also been found recently in reactive polymerization systems [85], supporting the plausibility of this *first AG hypothesis* that the barrier height $\mathcal{E}$ increases with the CRR "polymerization index" z^* [86].

The final link between this generalized Arrhenius description for the precipitous increase of relaxation times in cooled liquids and GD theory of glass formation is achieved through a second, less evident, hypothesis of the AG model that z^* is simply inversely proportional to the configurational entropy [87]. This relation is partially motivated by earlier observations by Bestul and Chang that the entropic barrier height $\mathcal{E}$ in a generalized Arrhenius relation varies nearly inversely [88, 89] to the fluid entropy S (relative to that of the glass) at low temperatures near T_g. Thus, the second AG hypothesis also has a reasonable basis, at least at a phenomenological level. As discussed later, the second AG hypothesis is consistent with the widely utilized relation for the temperature-dependent activation energy of Vogel–Fulcher–Tammann–Hesse (VFTH) over a wide temperature range near T_g, perhaps providing the strongest argument in favor of the second AG hypothesis [2, 3, 6, 7, 10].

Despite the heuristic nature of the reasoning involved, the AG model has held up remarkably well over the last 40 years in comparison with numerous experiments [25, 49, 90, 91] and with recent simulations of diverse glass-forming fluids [64–68, 92] (e.g., silica, binary Lennard-Jones mixtures, water, orthoterphenyl). Recent studies by Mohanty et al. [87], Lubchenko and Wolynes [93], and Bouchaud and Biroli [45] have sought to place the AG model on a sounder theoretical foundation. Good agreement with the AG model has been reported for simulations performed at temperatures well above T_g, where the configurational entropy is estimated through the "energy landscape" or "inherent structure" construction of Stillinger and Weber [94], rather than from the fluid's excess entropy S_{exc}, which is determined from specific heat measurements as the difference between the fluid entropy S and that of the crystal or glass [95]. On the other hand, deviations from the AG relation have been claimed [49] in real glass-forming liquids at temperatures 20–30 K above T_g. The basis for these claims of a failure of the AG model at elevated temperatures is explained in Section IV.A, and a reformulation of the AG model in the terms of the entropy density is shown below to remove these difficulties.

III. THEORETICAL BACKGROUND

The lattice cluster theory [50, 51] (LCT) is a generalization of the FH mean field approximation to include the influence of monomer structure and local correlations on the thermodynamic properties of polymers in bulk. The theory is based on the use of an extended lattice model in which single site occupancy constraints still apply, but individual monomers cover several lattice sites to reflect the size and bonding patterns of the actual molecules. For example, Fig. 3a depicts several structured monomers for polymer species that have been treated by our lattice model of polymers. The LCT generates an approximation for the free energy of polymer systems by employing a double expansion in the reciprocal of the lattice coordination number z and in the dimensionless van der Waals interaction energies $\{\epsilon_{i,j}\}/k_B T$, with the FH free energy as the zeroth order term. The LCT free energy expression includes contributions from short range correlations that arise from polymer chain connectivity, van der Waals interactions, chain semiflexibility, and monomer structure and that are absent in FH theory, which cannot distinguish between the monomer structures in Fig. 3a. Moreover, this free energy is obtained analytically, with the monomer structure described in terms of readily evaluated geometrical indices $\{N_\alpha\}$ that introduce the distinction between different monomer structures and polymer chain architectures.

Before discussing the LCT in detail, we note that the theory provides a description of polymer fluids and does not admit the possibility that a solid forms through crystallization. Hence, the term *equilibrium* in this chapter designates an *equilibrated fluid* for which the apparent thermodynamic properties are independent of time in practical measurements. The melting temperature T_m for our model is not known in general, but below T_m the supercooled liquid exists in a metastable state with respect to the true equilibrium crystal. As mentioned earlier, we expect T_A often to approximate T_m in fluids that crystallize, so that this metastable regime often corresponds to the temperature range in which polymers are processed. These potentially metastable extensions of the liquid state have physical significance for the observable properties of many polymers, where the crystalline state is physically inaccessible due to extremely high kinetic barriers. Thus, while polymer fluids may not be in strict thermodynamic equilibrium under the

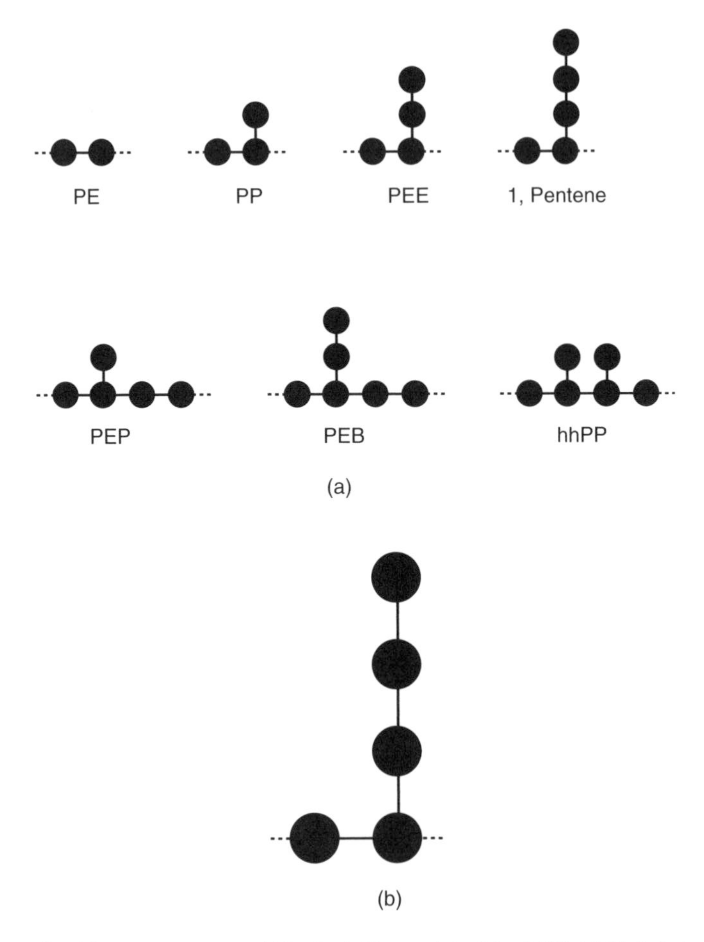

Figure 3. Monomer structure within the lattice model of the LCT. (a) United atom group monomer structures of several polyolefins. Circles denote CH_n ($n = 0$ to 3) groups, solid lines represent the C—C bonds between these groups, and dotted lines indicate the groups that are linked to the neighboring monomers in the polyolefin chains. (b) United atom group monomer structure chosen for the schematic model of glass formation. Circles denote CH_n groups, solid lines designate chemical bonds between these groups, and dashed lines indicate the groups that are linked to the CH_n units of the neighboring monomers in the chain backbone.

conditions in which they are processed and utilized, they are often in "metastable equilibrium"(provided that crystallization can in principle occur), and their properties can thus be reasonably described by thermodynamics in the temperature range $T_g < T < T_m$. Of course, our theory ceases to apply if crystallization begins to occur in the polymer fluid, a situation that can significantly affect the properties (e.g., fragility) of the fluid.

A. Lattice Cluster Theory of Glass Formation in Polymer Melts

The lattice cluster theory (LCT) for glass formation in polymers focuses on the evaluation of the system's configurational entropy $S_c(T)$. Following Gibbs–DiMarzio theory [47, 60], S_c is defined in terms of the logarithm of the microcanonical ensemble (fixed N, V, and U) density of states $\Omega(U)$,

$$S_c(T) = k_B \ln \Omega(U)|_{U=U(T)} \tag{1}$$

where $U(T)$ is the internal energy at temperature T and k_B is Boltzmann's constant. By definition, $S_c(T)$ must be a monotonically increasing function of T.

The configurational entropy per lattice site s_c, is related to the total (microcanonical) entropy S_c as

$$s_c = S_c/N_l \tag{2}$$

where N_l is the total number of lattice sites, which is proportional to the system volume V. The quantity s_c is an "entropy density" S_c/V, which need not be monotonic in T.

The density of states $\Omega(U)$, in turn, is connected with the canonical ensemble partition function $Z(T)$ through the formal weighted sum over states,

$$Z(T) = \int_0^\infty dU \Omega(U) \exp(-U/k_B T) \tag{3}$$

The computation of the density of states within the LCT is nontrivial since the LCT free energy $F(T) = -\beta \ln Z(T)$ is derived as an expansion in the product of the van der Waals interaction energies $\{\epsilon\}$ and $\beta = 1/k_B T$. This free energy series for semiflexible polymer systems has been developed [50, 51] through second order in β,

$$\beta F/N_l = A + B\beta - C\beta^2 \tag{4}$$

where the coefficients A, B, and C are functions (for given z) of the volume fractions and molecular masses of the system's components, the van der Waals interaction energies $\{\epsilon\}$, the dimensionless conformational energy differences $\beta\{E\}$ between *trans* and *gauche* conformations (i.e., the chain stiffness), as well as a set of geometrical indices that reflect the chemical structure of the system's constituents. The "entropic" terms appear in the coefficient A, but A depends on temperature through the $\beta\{E\}$. Since the LCT has been formulated for hypercubic lattices ($z = 2d$), the *trans* conformation is defined as that for a pair of collinear successive bonds, while the *gauche* conformation arises when the two successive bonds lie along orthogonal directions [51]. The inclusion of energy differences between *trans* and *gauche* conformations into the theory, that is, the treatment of the polymers as semiflexible, greatly enhances the complexity of the evaluation of $s_c(T)$ because A, B, and C of Eq. (4) then become functions of $\{\exp(-\beta E)\}$.

The truncation of the high temperature series in Eq. (4) at order β^2 is a valid concern when applying the theory at low temperatures. This concern also extends to GD theory, which implicitly involves a truncation at order β. At some point, these perturbative treatments must simply fail, but we expect the lattice theories to identify faithfully the location of the entropy crisis at low temperatures, based on numerous previous comparisons between measurements and GD theory. Experience [96] with the LCT in describing equation of state [97] and miscibility [98] data indicates that this approach gives sensible and often accurate estimates of thermodynamic properties over wide ranges of temperatures and pressures. In light of these limitations, we focus on the temperature range above T_g, where the theory is more reliable.

1. One-Bending Energy Model

The individual monomers are represented in terms of a set of united atom groups that each occupies a single lattice site, so that an individual monomer spreads out over several neighboring lattice sites to reflect its size and shape. All united atom groups are assumed, for simplicity, to interact with a common monomer averaged van der Waals interaction energy ϵ, and a common *gauche* energy penalty E is ascribed to all *gauche* semiflexible bond pairs, regardless of whether the *gauche* bonds belong to the backbone or to side chains. The description of a one component monodisperse polymer system involves the use of only one composition variable ϕ and one site occupancy index M (proportional to the molar mass M_{mol}), which is defined as the number of united atom groups in a single chain. Under these general conditions, the functional dependence of the coefficients A, B, and C of Eq. (4) can be formally written as

$$A(E) = A_o + A_1(E) = A_o(\phi, M, N_{2i}) + A_1(\phi, M, \{N_\alpha\}, \epsilon, g_E) \tag{5}$$

$$B(E) = B(\phi, M, \{N_\alpha\}, \epsilon, g_E) \tag{6}$$

$$C(E) = C(\phi, M, \{N_\alpha\}, \epsilon, g_E) \tag{7}$$

where A_o and $A_1(E)$ represent the athermal and nonathermal portions of $A(E)$, respectively, where $\{N_\alpha\}$ are geometrical indices that specify a given monomer's structure, where g_E is the bending energy factor [50, 51],

$$g_E = \frac{q\exp(-\beta E)}{1 + (q-1)\exp(-\beta E)} \tag{8}$$

and where q is the total number of *trans* and *gauche* conformations for a given monomer species. The factor g_E plays the role of the "order parameter" for chain semiflexibility. Explicit formulas for A, B, and C for multicomponent polymer systems are given in Ref. 50 (with corrections in Appendix A of Ref. 52), and the geometrical indices are described and tabulated in Refs. 52 and 53. Note that q is simply a measure of the orientational entropy associated with a consecutive pair of completely flexible bonds. The form of Eq. (8) is obtained by assuming the presence of one *trans* and $(q-1)$ *gauche* conformations for a pair of successive bonds between united atom groups. If $E = 0$, the bonds are fully flexible, whereas the bonds are completely rigid in the $E \to \infty$ limit. The quantity A_o of Eq. (5) contains the combinatorial portion (see below) of the dimensionless free energy $\beta F/N_1$, and N_{2i} designates the number of bond pairs in a single chain that can be assigned as being either in *trans* or *gauche* conformations [51].

Given the above model and assumptions, Freed [50] has recently shown that the LCT configurational entropy $s_c(T)$ of a polymer melt is given by the expression

$$\frac{s_c(T)}{k_B} = -\left[\mathcal{A}_o(f=f_o) + \mathcal{A}_1(f=f_o) + \frac{[\beta\mathcal{B}(f=f_o) - \beta u(T) + \beta E f_o N_{2i}/M]^2}{4\beta^2\mathcal{C}(f=f_o)}\right] \tag{9}$$

where f and f_o are defined below. In Eq. (9), $u(T)$ is the specific internal energy, which is derived from Eq. (4) as

$$\beta u \equiv \beta U(T)/N_1 = \beta B - 2\beta^2 C - \frac{\beta E g_E^2}{q\exp(-\beta E)}\frac{\partial}{\partial g_E}(A + \beta B - \beta^2 C) \tag{10}$$

$\mathcal{A}_o(f)$ is equal to

$$\begin{aligned}\mathcal{A}_o(f) = {} & \frac{\phi}{M}\ln\left(\frac{2\phi}{z^l M}\right) + \phi\left(1 - \frac{1}{M}\right) + (1-\phi)\ln(1-\phi) \\ & + \phi\frac{N_{2i}}{M}[-f\ln(q-1) + f\ln f + (1-f)\ln(1-f)]\end{aligned} \tag{11}$$

and $\mathcal{A}_1(f)$, $\mathcal{B}(f)$, and $\mathcal{C}(f)$ are generalizations of A_1, B, and C in Eqs. (5)–(7), obtained by replacing the bending energy factor g_E from Eq. (8) with the variable [50]

$$g_f = f\frac{q}{q-1} \tag{12}$$

The quantity f in Eq. (12) represents the ratio of the total number of *gauche* bond pairs to the maximum number of *gauche* bond pairs (N_{2i} per chain) in the system; l in Eq. (11) designates the number of subchains in a single chain; and z is the lattice coordination number ($z = 2d = 6$ for a simple cubic lattice). Each subchain is defined as a combination of successive bonds that may reside in either *trans* or *gauche* conformations [51]. A *gauche* energy penalty E is not assigned to successive bonds that belong to different subchains [50, 51]. The quantities $\mathcal{A}_1$, $\mathcal{B}$, and $\mathcal{C}$ are analogues of Eqs. (5)–(7) that are defined in terms of f,

$$\mathcal{A}_1(f) = A_1(\phi, M, \{N_\alpha\}, \epsilon, g_f) \tag{13}$$

$$\mathcal{B}(f) = B(\phi, M, \{N_\alpha\}, \epsilon, g_f) \tag{14}$$

$$\mathcal{C}(f) = C(\phi, M, \{N_\alpha\}, \epsilon, g_f) \tag{15}$$

Properties computed in terms of the fraction f of *gauche* bonds are obtained within a constrained ensemble that we term the "f-ensemble," while those computed in terms of g_E are said to be derived from the "E-ensemble."

The f_o that appears in Eq. (9) is determined as the value of f that maximizes the configurational entropy $s_c(T)$, that is, from the condition

$$\frac{\partial}{\partial f}\left[\mathcal{A}_o(f) + \mathcal{A}_1(f) + \frac{[\beta\mathcal{B}(f) - \beta u + \beta E f N_{2i}/M]^2}{4\beta^2\mathcal{C}(f)}\right]\Bigg|_{\phi,T} = 0 \tag{16}$$

While Eqs. (9)–(16) provide a recipe for evaluating $s_c(T)$ for constant volume (V) systems (i.e., constant ϕ), they can easily be applied to constant pressure systems by computing ϕ for a given pressure P and temperature T from the equation of state. For internal consistency, this equation of state must be derived from the free energy $\mathcal{F}$ expression appropriate to the f-ensemble,

$$P = -\frac{\partial\mathcal{F}(f_o)}{\partial V}\bigg|_{n,T} = -\frac{1}{a_{\text{cell}}^3}\frac{\partial\mathcal{F}(f_o)}{\partial N_1}\bigg|_{n,T} \tag{17}$$

where n is the total number of polymer chains, a_{cell}^3 is the volume associated with a single lattice site, and the free energy $\mathcal{F}(f)$ is given by

$$\beta\mathcal{F}(f) = N_l[\mathcal{A}_o(f) + \mathcal{A}_1(f) + \beta\mathcal{B}(f) - \beta^2\mathcal{C}(f)] \tag{18}$$

Other thermodynamic properties are similarly evaluated from $\mathcal{F}(f_o)$.

2. *Two-Bending Energy Model*

The above mean field treatment of s_c represents an extension of GD theory to a melt of semiflexible, interacting polymers composed of structured monomers. Equations (9)–(11) and (16)–(18) maintain their general validity when different united atom groups within the individual monomers interact with different van der Waals energies, but they are *restricted* to models with a single bending energy E for the whole molecule. While the use of a common *gauche* energy penalty E for all bond pairs is a natural point of departure, this assumption is clearly an oversimplification for treating many real polymers, such as polystyrene, where the side groups are rather rigid compared to the backbone. Since the different rigidities of the backbone and side groups are expected to affect the strength of the temperature dependence of $s_c(T)$ (and thus the fragility of the glass formation process [77–79]), we extend the theory to describe a model polymer melt in which the *gauche* energy penalty E_b in the chain backbone *differs* from the *gauche* energy penalty E_s in the side groups. The existence of two bending energies E_b and E_s leads to the appearance of two separate *gauche* bond fractions f_b and f_s in the theory, where the subscripts b and s indicate backbone and side chains, respectively. The extension of the theory also implies a partitioning of the geometrical indices $\{N_\alpha\}$ (defined per chain) into three groups $\{N_\alpha^{(b)}\}$, $\{N_\alpha^{(s)}\}$, and $\{N_\alpha^{(bs)}\}$. These three groups refer to classes of configurations with pairs, triads, etc., of semiflexible bonds belonging to the chain backbone, the side group, and both, and these classes are denoted, respectively, by the superscripts (b), (s), and (bs). (For instance, $N_{2i} = N_{2i}^{(b)} + N_{2i}^{(s)}$ and $N_{4+} = N_{4+}^{(s)} + N_{4+}^{(bs)}$, where N_{2i} is the number of sets of two consecutive bonds that lie along one subchain and N_{4+} is the number of sets of four bonds that meet at a common united atom group [50].) Consequently, there are two bending energy factors g_{E_b} and g_{E_s} defined as

$$g_{E_b} = \frac{q_b \exp(-\beta E_b)}{1 + (q_b - 1)\exp(-\beta E_b)}, \qquad g_{E_s} = \frac{q_s \exp(-\beta E_s)}{1 + (q_s - 1)\exp(-\beta E_s)} \tag{19}$$

where q_b and q_s denote total numbers of *trans* and *gauche* conformations for a pair of consecutive bonds in the backbone and side chains, respectively. In general, q_b and q_s may differ, but we assume a single *trans* conformation and two *gauche* conformations for both the backbone and side groups.

The configurational entropy $s_c(T)$ for the two-bending energy model is expressed in the form

$$\frac{s_c(T)}{k_B} = -\mathcal{A}_o(f_b = f_b^o, f_s = f_s^o) - \mathcal{A}_1(f_b = f_b^o, f_s = f_s^o) - \frac{[\beta\mathcal{B}(f_b = f_b^o, f_s = f_s^o) - \beta u(T) + \beta E_b f_b^o N_{2i}^{(b)}/M + \beta E_s f_s^o N_{2i}^{(s)}/M]^2}{4\beta^2 \mathcal{C}(f_b = f_b^o, f_s = f_s^o)} \tag{20}$$

where

$$\mathcal{A}_o(f_b, f_s) = \frac{\phi}{M}\ln\left(\frac{2\phi}{z^l M}\right) + \phi\left(1 - \frac{1}{M}\right) + (1-\phi)\ln(1-\phi) + \phi\frac{N_{2i}^{(b)}}{M}[-f_b\ln(q_b - 1) + f_b\ln f_b + (1-f_b)\ln(1-f_b)] + \phi\frac{N_{2i}^{(s)}}{M}[-f_s\ln(q_s - 1) + f_s\ln f_s + (1-f_s)\ln(1-f_s)] \tag{21}$$

and the polynomials,

$$\mathcal{A}_1(f_b, f_s) = \mathcal{A}_1(\phi, M, \{N_\alpha^{(b)}\}, \{N_\alpha^{(s)}\}, \{N_\alpha^{(bs)}\}, \epsilon, g_{f_b}, g_{f_s}) \tag{22}$$

$$\mathcal{B}(f_b, f_s) = \mathcal{B}(\phi, M, \{N_\alpha^{(b)}\}, \{N_\alpha^{(s)}\}, \{N_\alpha^{(bs)}\}, \epsilon, g_{f_b}, g_{f_s}) \tag{23}$$

$$\mathcal{C}(f_b, f_s) = \mathcal{C}(\phi, M, \{N_\alpha^{(b)}\}, \{N_\alpha^{(s)}\}, \{N_\alpha^{(bs)}\}, \epsilon, g_{f_b}, g_{f_s}) \tag{24}$$

with

$$g_{f_b} = f_b\frac{q_b}{q_b - 1}, \quad g_{f_s} = f_s\frac{q_s}{q_s - 1} \tag{25}$$

are generated from Eqs. (13)–(15) by partitioning the geometrical factors $\{N_\alpha\}$ into three groups $\{N_\alpha^{(b)}\}$, $\{N_\alpha^{(s)}\}$, and $\{N_\alpha^{(bs)}\}$ and by multiplying them by the corresponding statistical weights involving, respectively, g_{f_b}, g_{f_s}, and $g_{f_b}g_{f_s}$ (see Appendix B of Ref. 52 for more details).

The specific internal energy βu of Eq. (20) is given by

$$\begin{aligned}\beta u = {} & \beta B(E_b, E_s) - 2\beta^2 C(E_b, E_s) \\ & - \frac{\beta E_b g_{E_b}^2}{q_b \exp(-\beta E_b)} \frac{\partial}{\partial g_{E_b}} [A(E_b, E_s) + \beta B(E_b, E_s) - \beta^2 C(E_b, E_s)] \\ & - \frac{\beta E_s g_{E_s}^2}{q_s \exp(-\beta E_s)} \frac{\partial}{\partial g_{E_s}} [A(E_b, E_s) + \beta B(E_b, E_s) - \beta^2 C(E_b, E_s)] \end{aligned} \tag{26}$$

where the functions $A(E_b, E_s) = A_o + A_1(E_b, E_s)$, $B(E_b, E_s)$, and $C(E_b, E_s)$ are the E-ensemble counterparts of $\mathcal{A}$, $\mathcal{B}$, and $\mathcal{C}$, respectively.

The values f_b^o and f_s^o of Eq. (20) are obtained by solving the following set of equations:

$$\frac{\partial}{\partial f_b}\left[\mathcal{A}_o(f_b, f_s) + \mathcal{A}_1(f_b, f_s) + \frac{[\beta\mathcal{B}(f_b, f_s) - \beta u + \beta E_b f_b N_{2i}^{(b)}/M + \beta E_s f_s N_{2i}^{(s)}/M]^2}{4\beta^2 \mathcal{C}(f_b, f_s)}\right]\Bigg|_{\phi, T} = 0 \tag{27}$$

$$\frac{\partial}{\partial f_s}\left[\mathcal{A}_o(f_b, f_s) + \mathcal{A}_1(f_b, f_s) + \frac{[\beta\mathcal{B}(f_b, f_s) - \beta u + \beta E_b f_b N_{2i}^{(b)}/M + \beta E_s f_s N_{2i}^{(s)}/M]^2}{4\beta^2 \mathcal{C}(f_b, f_s)}\right]\Bigg|_{\phi, T} = 0 \tag{28}$$

In analogy to Eq. (17), the equation of state is derived from the corresponding free energy expression in the f-ensemble,

$$\beta\mathcal{F}(f_b, f_s) = N_l[\mathcal{A}_o(f_b, f_s) + \mathcal{A}_1(f_b, f_s) + \beta\mathcal{B}(f_b, f_s) - \beta^2\mathcal{C}(f_b, f_s)] \tag{29}$$

by taking the appropriate derivative,

$$P = -\frac{\partial \mathcal{F}(f_b^o, f_s^o)}{\partial V}\bigg|_{n,T} = -\frac{1}{a_{\text{cell}}^3} \frac{\partial \mathcal{F}(f_b^o, f_s^o)}{\partial N_l}\bigg|_{n,T} \tag{30}$$

The generalization of Eqs. (19)–(30) to a polymer melt with arbitrary numbers of distinct bending energies in both backbone and side chains is straightforward.

The equation of state $P = P(T, \phi)$ enables the computation of the isothermal compressibility κ_T,

$$\kappa_T = -\frac{1}{V}\frac{\partial V}{\partial P}\bigg|_T \tag{31}$$

and the specific volume v,

$$v(T,P) = \frac{1}{\phi}\frac{a_{\text{cell}}^3 r N_{\text{Av}}}{M_{\text{mol}}} \tag{32}$$

where r denotes the number of lattice sites occupied by a single monomer, M_{mol} is its molar mass, and N_{Av} is Avogadro's number. Both κ_T and $v(T,P)$ play an important role in analyzing glass formation [53, 54]. All other thermodynamic properties at constant T and P may readily be evaluated from $\mathcal{F}$ of Eq. (29). This two-bending energy model is considered here as a general schematic model (SM) for glass formation in polymeric liquids, based on the physical motivation described in the next subsection.

B. General Classes of Glass-Forming Polymers

Experimental studies of the structural origin of fragility in polymers suggest that polymers with simple flexible side branch structures are rather strong glass formers, while polymers with bulky, rigid side groups are more fragile [77, 78]. This empirically discovered idealized view that polymers can be classified [78] approximately into three basic categories—chains with a flexible backbone and flexible side groups, chains that have a flexible backbone and rigid side branches, and chains with a relatively stiff backbone and flexible side groups—provides a motivation for using the two-bending energy version of the LCT in our theoretical studies of the nature of glass formation in polymer melts. We term these categories as the flexible–flexible (F-F), flexible–stiff (F-S), and stiff-flexible (S–F) class polymers, respectively. This idealized classification ignores variability in the degree of bond rigidity (or flexibility) within both the backbone and side chains. The F-F polymers are an idealization of polydimethylsiloxane (PDMS), polyisobutylene (PIB), polyisoprene (PI), polyethylene (PE), and other species having both flexible backbone and side groups. The F-S polymers of the second class are intended to model molecules, such as polystyrene and poly(α-methylstyrene), where the side groups are spatially extended and relatively stiff on average, compared to the bonds in the chain backbone. Finally, the S-F chains describe polymers, such as poly(n-alkyl methacrylates), polycarbonates, and other polymers for which the chain backbone is comprised of aromatic or other rigid bulky groups. Polymers of these three general classes can be schematically represented by chains having different bending energies E_{b} and E_{s}. As shown in Fig. 3b, a single monomer in the calculations for our schematic model of glass-forming polymers is composed of five united atom groups; two of them are located in the chain backbone and the remaining three reside in the side chain. This simplified model contains a minimal set of physical parameters (E_{b}, E_{s}, and ϵ) and has been chosen because our goal here is to elucidate the universal characteristics of these broad classes of polymers, rather than to consider specific polymer fluids.

As mentioned earlier, the only difference between the F-F, F-S, and S-F polymers in our schematic model is the relative magnitude of the side group bending energy E_s and the chain backbone bending energy E_b. The class of F-F polymers is modeled by taking E_b/k_B and E_s/k_B to be identical and equal to the representative value of 400 K. The same value $E_b/k_B = 400$ K is ascribed to the F-S chains, but a relatively large $E_s/k_B = 4000$ K is used to model stiff side groups. The S-F case, where E_b and E_s are selected as $E_b/k_B = 700$ K and $E_s/k_B = 200$ K, is briefly analyzed in Section VIII. As explained later, this choice of E_b and E_s is designed to enable comparison with observed values of T_g in a series of poly(n-alkyl methacrylates) [99]. Figures 4a and 4b display the conformations of the F-F and F-S polymers that have been generated by MD simulations for *isolated* chains. The simulated configurations [100] are intended here only to convey a qualitative picture of how the relative rigidities of the side groups and the chain backbone govern the molecular architecture of the chain.

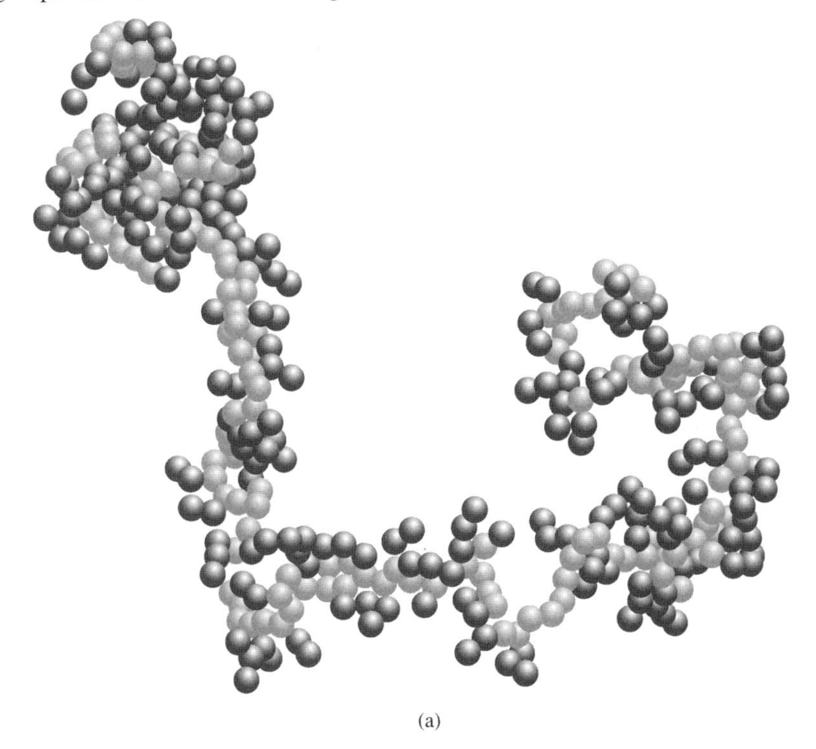

(a)

Figure 4. Representative chain configurations for F-F (a) and F-S (b) polymer classes. MD simulations [100] are obtained for isolated chains and are only meant for illustrative purposes. The polymerization index N is taken as $N = 100$. The monomer structure is assumed to be common for these two polymer categories and is depicted in Fig. 3b. The bending energies E_b and E_s are chosen in the simulations as $E_b/k_BT = E_s/k_BT = 0$, and $E_b/k_BT = 0$, $E_s/k_BT = 200$ for F-F and F-S polymer classes, respectively.

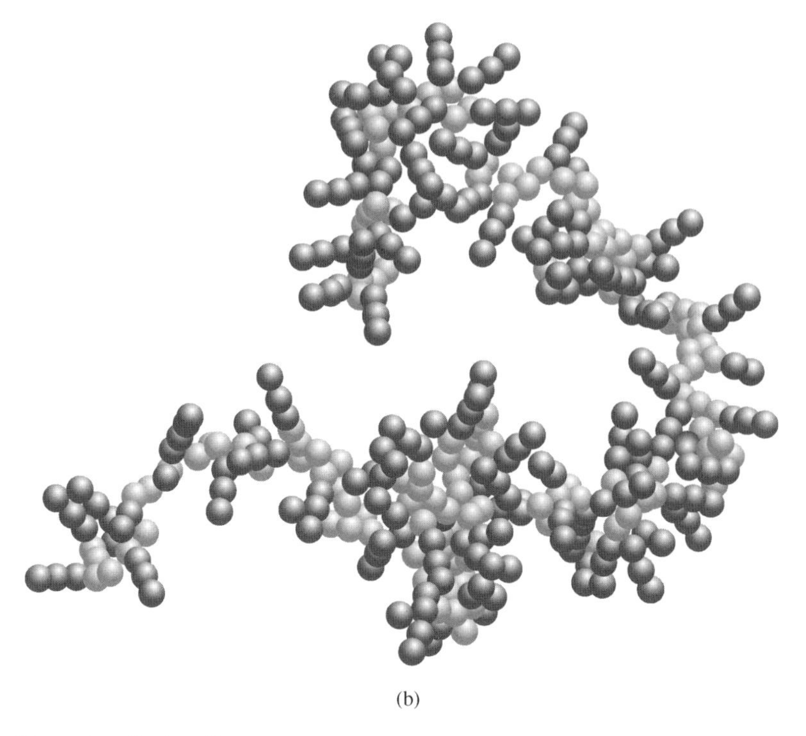

(b)

Figure 4. *(Continued)*

All computations apply for a pressure $P = 1$ atm (0.101325 MPa), unless otherwise specified, and have been conducted by choosing the nearest neighbor van der Waals interaction energy as $\epsilon/k_B = 200$ K (a typical value [101] for poly(α-olefins)) and the unit volume $v_{cell} = a_{cell}^3 = (2.7)^3 \mathring{A}^3$. We further assume the presence of one *trans* and two *gauche* conformations for each backbone and side chain bond pairs (i.e., $q_b = q_s = 3$). For the above choice of q_b, q_s, E_b, E_s, ϵ, and v_{cell}, the calculated $T_g(P = 1$ atm) for F-F polymers is about 250 K, which is a typical order of magnitude for T_g of flexible chain polymers with flexible side groups. A similar consistency is also obtained for the F-S polymers, where the computed T_g for $P = 1$ atm is about 350 K and is close to the T_g of polymers with rigid side groups.

C. Adam–Gibbs Relation Between the Configurational Entropy s_c and Relaxation in Glass-Forming Liquids

As discussed in Section II, the Adam–Gibbs [48] model of relaxation in cooled liquids relates the structural relaxation times τ, associated with long wavelength relaxation processes (viscosity, translational diffusion, rates of diffusion-limited chemical reactions, etc.), to the configurational entropy s_c through the generalized Arrhenius relation,

$$\tau = \tau_o \exp(\beta \mathcal{E}_{AG}), \quad \mathcal{E}_{AG} \equiv \Delta\mu[s_c^*/s_c(T)] \tag{33}$$

where τ_o is the high temperature limiting relaxation time τ in the fluid, $\Delta\mu$ is a (property and system dependent) activation energy at high enough temperatures, such that an Arrhenius dependence of τ approximately holds (i.e, $\tau = \tau_o \exp(\beta\Delta\mu)$), and s_c^* is the (assumed constant) high temperature limit of $s_c(T)$. For simple atomic fluids, we expect $\tau_o \sim \mathcal{O}(10^{-14}\ \mathrm{s})$, while τ_o may be somewhat larger ($\tau_o \sim \mathcal{O}(10^{-13}\ \mathrm{s})$) for more complex molecules [102] such as polymers, since the mobile fluid elements (monomers) are larger. The average activation energy $\mathcal{E}_{AG}$ in Eq. (33) grows upon cooling as particle motion becomes more collective. The ratio $z^* \equiv s_c^*/s_c(T)$ provides a measure of the number of particles in the dynamic clusters (cooperative rearranging regions or CRRs) embodying this collective motion. In the AG theory, the ratio z^* describes the enhancement of the activation energy ($\mathcal{E}_{AG}$) relative to its high temperature value $\Delta\mu$ (see Eq. (33)).

The experimental inaccessibility of the configurational entropy poses no problem for the LCT, apart from a consideration of whether to normalize the configurational entropy per lattice site or per monomer in order to provide a better representation of experiment within the AG model. Once the appropriate normalization of s_c has been identified, τ can be calculated from Eq. (33) as a function of temperature T, molar mass M_{mol}, pressure P, monomer structure, backbone and side group rigidities, and so on, *provided* that $\Delta\mu$ is specified [54]. The direct determination of $\Delta\mu$ from data for $T > T_A$ is not possible for polymer systems because T_A generally exceeds the decomposition temperature for these systems. Section V reviews available information that enables specifying $\Delta\mu$ for polymer melts.

Unfortunately, reliable experimental estimates of the configurational entropy are unavailable to enable explicit application of the AG model for polymer fluids. Instead, the temperature dependence of τ in polymer melts is often analyzed in terms of the empirical Vogel–Fulcher–Tammann–Hesse (VFTH) equation [103],

$$\tau = \tau_{\mathrm{VFTH}} \exp\left(\frac{DT_\infty}{T - T_\infty}\right) \tag{34}$$

where T_∞ is the "Vogel temperature" at which structural relaxation times τ and the shear viscosity η extrapolate to infinity, D is a "fragility constant" [104, 105] describing the strength of the temperature dependence of τ, and τ_{VFTH} is an adjustable parameter. In recent literature [92], the inverse of D has been advocated as a more suitable definition of fragility since $K \equiv 1/D$ is larger for more fragile fluids. Equation (34) is consistent with Eq. (33) if $s_c T$ varies linearly near T_0 as $s_c T \sim (T - T_0)$, and below we specify the conditions under which this behavior holds.

IV. CONFIGURATIONAL ENTROPY AND CHARACTERISTIC TEMPERATURES FOR GLASS FORMATION IN POLYMERIC FLUIDS

According to the classical GD theory of glass formation [47], an "ideal" glass transition can be identified with a thermodynamic event, the vanishing of the configurational entropy. This is a physically natural condition because a system must have a multiplicity of accessible configurational states to achieve equilibrium and exhibit relaxation. In analogy to the GD theory, our LCT configurational entropy s_c also extrapolates to zero at the temperature T_0. We are cognizant, however, that the literal interpretation of a vanishing s_c is uncertain because of the truncation of the high temperature expansion in Eq. (4) and because recent simulation studies suggest that s_c remains positive at low temperatures [73]. As discussed in Section II, we adopt a modified view of T_0 as the temperature at which the *excess* configurational entropy relative to the entropy of the glass extrapolates to zero, reflecting the sparseness of accessible configurational states. Thus, T_0 can be identified with the VFTH temperature T_∞ at which structural relaxation times τ *extrapolate* to infinity, in agreement with the AG concept of glass formation. As mentioned before, T_0 is not generally equal [106, 107] to the Kauzmann temperature T_K at which the excess molar entropy S_{exc} extrapolates to zero, but T_K and T_∞ are often found to be close to each other [91].

This section describes LCT calculations for the specific configurational entropy $s_c(T)$ for two limiting models of polymers: flexible chains with flexible side groups (F-F class) and flexible chains with relatively stiff side groups (F-S class). Experimental studies indicate that smooth, compact, symmetrical chains exhibit "strong" glass formation, while fragile polymers have more rigid backbones and sterically hindered pendant groups [77–79]. We then seek to determine if these recognized trends in fragility can be comprehended within the LCT and the "schematic model" of glass formation introduced in the previous section.

A. Temperature Dependence of the Configurational Entropy

The configurational entropy per occupied lattice site (i.e., per unit mass rather than per unit volume) is *by definition* a monotonic function of temperature, and, of course, the fluid entropy deduced from calorimetric measurements also has this monotonic property. Figure 5a compares the mass and site configurational entropies computed from the LCT as functions of temperature T for the F-F and

F-S polymer classes. The two types of configurational entropy nearly coincide with each other for temperatures lower than T_I (defined in next paragraph), but they differ appreciably at higher temperatures. The bifurcation in the configurational entropies in Fig. 5a is remarkably similar in form to the deviation observed between the experimentally determined excess fluid entropy $S_{exc}^{(mol)}$ and the configurational entropy as estimated from fits of relaxation data to the AG relation for τ (see Fig. 5b, c) [49].

More specifically, Richert and Angell [49] sought to assess the accuracy of the AG relation by fitting its basic parameters to precise dielectric relaxation data over a wide range of temperatures. They find that an assumed AG expression for the relaxation time τ of model glass-forming fluids (salol and 2-methyl tetrahydrofuran) represents the data well, provided that the fitted configurational entropy has a *maximum* at high temperatures (occurring at $1.6T_K$ and $1.7T_K$, respectively). Inspection of the Richert and Angell data in Fig. 5b, c clearly indicates that estimates of the "configurational entropy" based on AG theory deviate qualitatively at high temperatures from the excess entropy S_{exc} obtained from specific heat measurements, and it is this deviation that is often

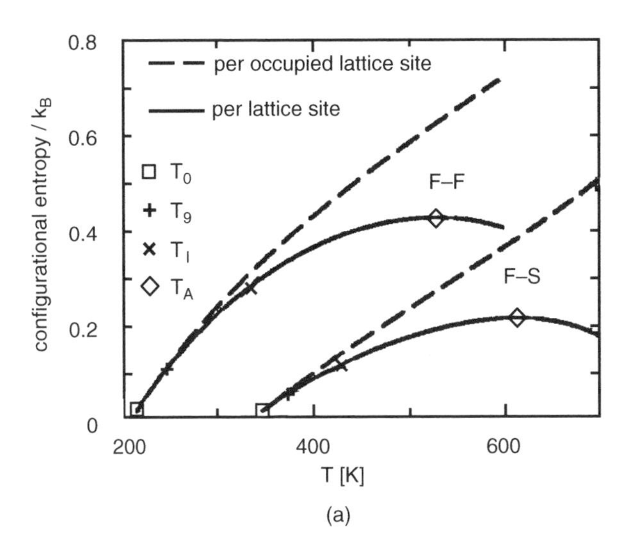

Figure 5. (a) Comparison of the LCT configuration entropy s_c (per site) with its counterpart $s_c^{(mass)}$ (per united atom group) for constant pressure ($P = 1$ atm; 0.101325 MPa) F-F and F-S polymer fluids ($M = 40001$). (Used with permission from J. Dudowicz, K. F. Freed, and J. F. Douglas, *Journal of Chemical Physics* **123**, 111102 (2005). Copyright © 2005, American Institute of Physics.) (b), (c) Comparison of the excess entropy $S_{exc}(T)$ (squares) estimated from specific heat measurements with the AG configurational entropy $S_c(T)$ (open circles) obtained from fits of the AG relation (33) to experimental dielectric relaxation data. (Used with permission from R. Richert and C. A. Angell, *Journal of Chemical Physics* **108**, 9016 (1998). Copyright © 1998, American Institute of Physics.)

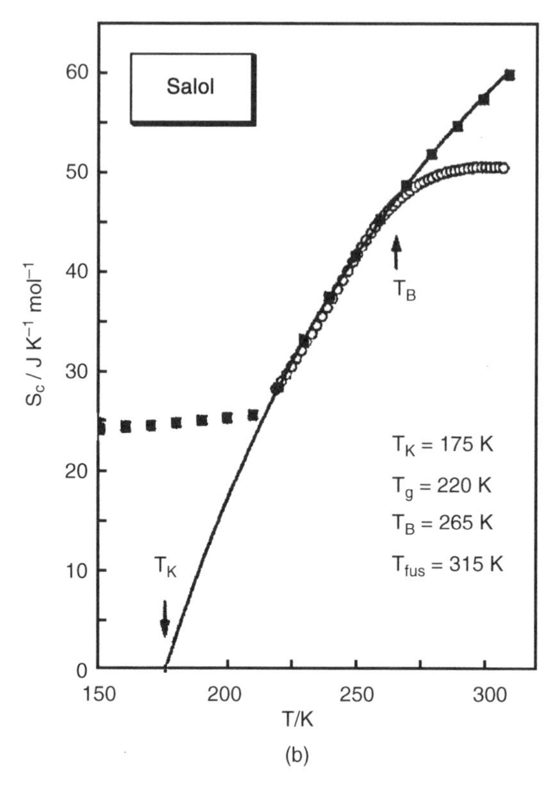

Figure 5. *(Continued)*

cited [15, 49] as evidence for a "breakdown" of the AG model. This conclusion, however, is difficult to reconcile with the fact that virtually all of the equilibrated simulation studies [55, 64–68, 92] indicating agreement with the AG model are restricted to relatively high temperatures where Richert and Angell [49] claim that the AG model fails!

The divergent trends between the excess fluid entropy S_{exc} and the AG based estimate of the configurational entropy from dielectric data also have their counterparts in comparisons between simulation estimates of the configurational entropy and S_{exc}. The configurational entropy determined by energy landscape calculations [64–68, 92] (denoted as $s_{c,L}$), as well as the nonvibrational fluid (site) entropy obtained from MC simulations [55, 73, 108], both tend to approach approximately *constant values* at high temperatures [109], while the molar excess entropy $S_{exc}^{(mol)}$ *does not* behave in this fashion, as made explicit from the measurements of Richert and Angell. Direct LCT calculation of the molar configurational entropy $S_c^{(mol)}$ for a constant pressure fluid of semiflexible polymers (see Fig. 5a) confirms that $S_c^{(mol)}$ *does not* saturate at high

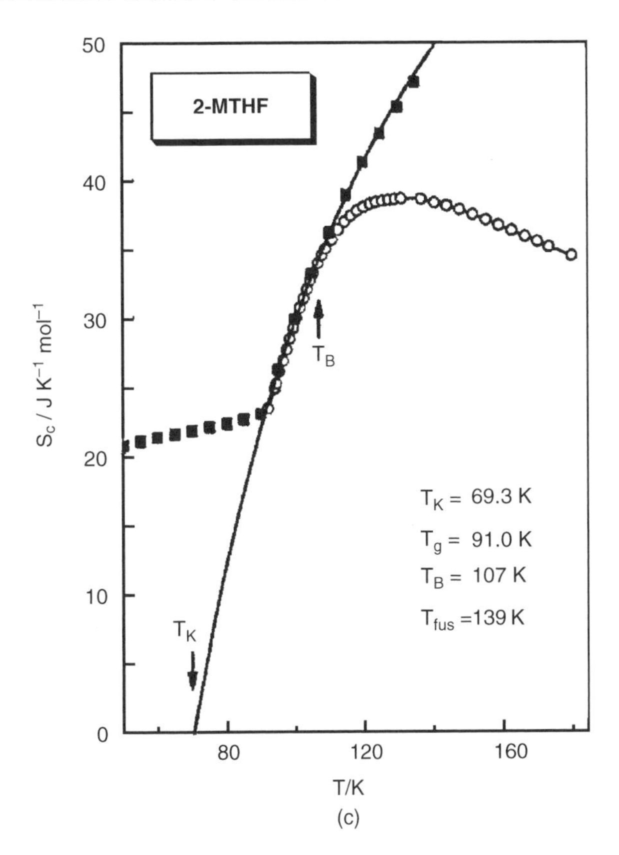

Figure 5. *(Continued)*

temperatures. In contrast, the computed entropy per site s_c varies similarly to trends found for the configurational entropy in the MC simulations of Wolfgardt et al. [73] and Kamath et al. [108] over the temperature ranges investigated.

The site entropy s_c is thus a sensible candidate for describing fluid relaxation outside the immediate vicinity of the glass transition. In a more precise language, s_c is actually an *entropy density*, and the maximum in $s_c(T)$ derives from an interplay between changes in the entropy and fluid density as the temperature is varied. Explicit calculations demonstrate that the maximum in $s_c(T)$ *disappears* in the limit of an incompressible fluid, which is physically achieved in the limit of infinite pressure. The pressure dependence of $s_c(T)$ is described in Section X, where it is found that the maximum in $s_c(T)$ becomes progressively shallower and T_A becomes larger with increasing pressure.

The configurational entropy of the AG model is identified by us with the entropy per lattice site s_c of the LCT, since the use of the entropy density is the only normalization condition that leads to a sensible variation of τ at high temperatures in the AG model (see Section IX). The experimental counterpart of the site entropy in the lattice model theory is the molar entropy $S_c^{(mol)}$ (or $S_{exc}^{(mol)}$) divided by the fluid molar volume $V^{(mol)}$ (superscripts (mol) denote molar quantities). (A similar idea of using an energy density in the description of the rate of structural relaxation in glassy materials has been introduced by Langer [110].) The breakdown of the AG theory reported by Richert and Angell may actually indicate that the molar excess fluid entropy is not really the appropriate representation of the configurational entropy in the AG model. While Binder et al. [55] have previously shown that that AG theory provides a good description of the temperature dependence of diffusion in polymer melts when $s(T)$ is identified with the site configurational entropy, they do not mention that this choice substantially departs from the use of S_{exc} in experimental tests of the AG theory under constant pressure conditions.

In addition to s_c vanishing at T_0, Fig. 5a shows that $s_c(T)$ exhibits a maximum s_c^* at a higher temperature T_A that is roughly *twice* T_0. As explained later, T_A is identified by us as the "Arrhenius temperature," above which structural relaxation times exhibit a nearly Arrhenius temperature dependence, $\tau \approx \exp(\beta\Delta\mu)$. Thus, T_A is the temperature below which collective motion initiates [111], that is, where $z^* \equiv s_c^*/s_c(T) > 1$. The LCT estimates of $s_c(T)$ for both the F-F and F-S polymers are displayed in Fig. 6 for relatively small and large representative M ($M = 101$ and $M = 40001$), where M is the number of united atom groups per chain, which is thus proportional to the molar mass M_{mol}. The entropy s_c in Fig. 6 is normalized by its maximum value s_c^* and is presented as a function of the reduced temperature δT,

$$\delta T \equiv (T - T_0)/T_0 \tag{35}$$

The inset to Fig. 6 exhibits s_c^* as depending sensitively on the polymer class, but relatively weakly on molar mass. The temperature variation of $s_c(T)/s_c^*$ is roughly linear for small δT near T_0, consistent with the empirical VFTH equation, as noted earlier. On the other hand, this dependence becomes roughly quadratic in δT at higher temperatures, where s_c achieves a maximum s_c^* at T_A. Attention in this chapter is primarily restricted to the broad temperature range ($T_0 < T < T_A$), where a decrease of s_c with T is expected to correspond to an increase in the extent of collective motion [112].

The two different temperature regimes of s_c occurring below T_A are separated from each other by a precisely determined crossover temperature T_I (denoted by crosses in Fig. 6), which is defined by an *inflection point* in the product $s_c(T)T$ as a function of T. As discussed later, this crossover or "inflection temperature" T_I appears to conform to the phenomenology of the experimentally estimated mode coupling temperature T_{mc}^{exp} (denoted also as T_c).

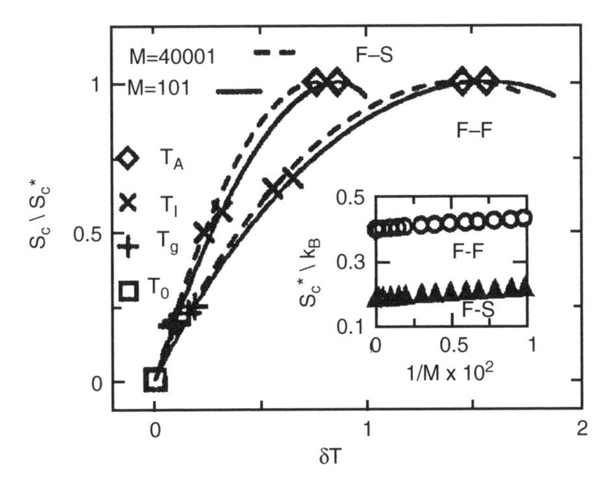

Figure 6. LCT configurational entropy s_c per lattice site of a constant pressure ($P = 1$ atm; 0.101325 MPa) polymer melt as a function of the reduced temperature $\delta T \equiv (T - T_0)/T_0$, defined relative to the ideal glass transition temperature T_0 at which s_c extrapolates to zero. The specific entropy s_c is normalized by its maximum value $s_c^* \equiv s_c(T = T_A)$ (see Fig. 5a). Different curves refer to the F-F and F-S classes of polymers (described in detail in the text) and to different numbers M of united atom groups in single chains. Note that M is proportional to the molar mass M_{mol} of the polymer fluid, and the values of $M = 101$ and $M = 40001$ correspond to low and high molar mass polymer chains, respectively. The characteristic temperature of glass formation, the ideal glass transition temperature T_0, the glass transition temperature T_g, the crossover temperature T_I, and the Arrhenius temperature T_A are indicated in the figure. The inset to Fig. 6 illustrates the $1/M$ dependence of the "critical site entropy" s_c^* for both F-F and F-S classes of polymers.

In summary, Fig. 6 exhibits the four characteristic temperatures of glass formation: the Arrhenius temperature T_A, the crossover temperature T_I, the ideal glass transition temperature T_0, and a kinetic glass-formation temperature T_g (defined in Section VI), to illustrate their relative locations with respect to the temperature variation of s_c.

Before further analyzing the implications of the temperature dependence of s_c on the nature of glass formation, we note that the present illustrative LCT calculations are primarily concerned with simple organic polymer fluids having predominantly carbon chain structures. The presence of heteroatoms (O, Cl, N, etc.) within the chain or side groups can alter the van der Waals interaction energy ϵ from the typical value ($\epsilon/k_B = 200$ K) that has been chosen for the calculations. Moreover, many polymers contain stiffer backbone segments than described by our representative choice of the backbone energy, $E_b/k_B = 400$ K, a choice that again derives from a preoccupation with typical synthetic polymers having simple atomic backbone structure (e.g., PIB). Additional calculations (not included here) demonstrate that all four characteristic temperatures T_0, T_g, T_I, and T_A increase roughly linearly with the van der Waals energy ϵ and the backbone bending energy E_b over a wide range of these energetic variables. (This linear variation is limited, of course, and saturation can occur for very large ϵ and E_b.) However, the general trends discussed here in the variation of these characteristic temperatures with molecular structure, pressure, or molar mass remain largely unchanged as ϵ and E_b are varied. Therefore, our assumption of specific values for ϵ, E_b, and E_s should not be overly restrictive in describing the general properties of the F-F and F-S polymer classes.

Closer inspection of Fig. 6 indicates that the temperature dependence of s_c/s_c^* is generally much weaker for F-F class polymers than for F-S class polymers. A weaker variation of s_c with temperature implies a smaller apparent temperature dependence of the activation energy $\mathcal{E}$ and, broadly speaking, a lower fragility. Thus, the F-S polymers have a greater fragility than the F-F chains, when a common activation energy $\Delta\mu$ is used for comparing these two cases, a finding that is consistent with the general trend identified in experimental studies of glass-forming fluids [77–79]. For a given reduced temperature $\delta T < (T_A - T_0)/T_0$, the ratio s_c/s_c^* (and therefore the fragility) slightly increases with M for both F-S and F-F polymer classes. The magnitude of the critical entropy s_c, that is, $s_c^* \equiv s_c(T = T_A)$, is not universal (see Table I), in contrast to the suggestion of AG [48, 63]. Indeed, s_c^* differs by nearly a factor of 2 between these two polymer classes, and the M dependence of the critical entropy s_c^* is found to be rather weak. The difference between the configurational entropies for the F-S and F-F classes of polymers is understandable, since a stiffening of the side chain branches naturally reduces the configurational entropy of the fluid, regardless of the temperature. The inset to Fig. 6 shows that this reduction in s_c^* between the F-F and F-S classes amounts to nearly a factor 2 (when the bending energy E_s is increased from $E_s/k_B = 400$ K to $E_s/k_B = 4000$ K) and that s_c^* is a nearly linear function of $1/M$. Johari [63] has recently emphasizes this nonuniversality of s_c^*, based on comparison of the AG model with experiments.

B. Temperature Dependence of the Size of the CRR

The ratio $z^* = s_c^*/s_c(T)$ in the AG model of structural relaxation is interpreted as the average number of monomer elements in the dynamic CRR structures that are hypothesized to form in cooled glass-forming liquids. (A substantial body of experimental [7] and simulation data [113–116] supporting the existence of these structures has been accumulated since the inception of the AG model [48, 117], providing many insights into the geometrical characteristics of the clusters.) The arguments [48] of AG specifically imply that the apparent activation energy $\mathcal{E}$ of cooled glass-forming fluids is simply the product of z^* and the magnitude of the high temperature limit of $\mathcal{E}$, that is, $\mathcal{E}_{AG}(T \geq T_A) \equiv \Delta\mu$, so that $\mathcal{E}_{AG}/\Delta\mu = z^*$. As emphasized by Kivelson and co-workers [13], the reduced activation energy $\mathcal{E}_{AG}$ is directly determined from transport properties (diffusion coefficients,

TABLE I
The Characteristic Parameters (s_c^*, z^*, C_s, K_s), Characteristic Temperatures (T_0, T_g, T_I, T_A), Characteristic Ratios (T_I/T_g, T_I/T_0, T_A/T_g, T_A/T_0, T_A/T_I, $\psi = T_I/(T_I - T_g)$), Specific Volume $v(T = T_g)$, Isothermal Compressibility κ_{T_g}, and Structural Relaxation Times τ as Computed from the Generalized Entropy Theory for the F-F and F-S Polymer Fluids at a Pressure of $P = 0.101325$ MPa (1 atm)

	F-F Polymer Fluid		F-S Polymer Fluid	
Property	$M = 101$	$M = 40001$	$M = 101$	$M = 40001$
s_c^*/k_B	0.437	0.405	0.222	0.193
$z^*(T = T_I)$	1.48	1.55	1.76	2.02
$z^*(T = T_g)$	4.07	4.46	5.10	5.53
C_s	2.79	2.87	6.53	7.10
K_s	0.181	0.200	0.308	0.361
T_0 (K)	200	216	314	349
T_g (K)	237	250	345	376
T_I (K)	329	336	414	430
T_A (K)	512	530	584	614
T_I/T_g	1.39	1.35	1.20	1.15
T_I/T_0	1.65	1.56	1.32	1.23
T_A/T_g	2.16	2.12	1.70	1.64
T_A/T_0	2.56	2.45	1.86	1.76
T_A/T_I	1.56	1.58	1.41	1.43
$\psi = T_I/(T_I - T_g)$	3.57	3.89	5.99	7.89
$\tau(T = T_A)$ (s)	$10^{-11.3}$	$10^{-11.3}$	$10^{-11.2}$	$10^{-11.2}$
$\tau(T = T_I)$ (s)	$10^{-9.15}$	$10^{-8.95}$	$10^{-8.41}$	$10^{-7.74}$
$\tau(T = T_g)$ (s)	$10^{1.74}$	$10^{2.66}$	$10^{2.96}$	$10^{3.51}$
$v(T = T_g)$ (m^3/kg)	862	868	931	958
κ_{T_g} (GPa^{-1})	0.330	0.384	1.26	1.68
$1/\kappa_{T_g}$ (GPa)	3.03	2.60	0.791	0.595

viscosity) without the need for assumptions concerning the vanishing of S_c or the nature of the glass-formation process. We next consider the ratio $z^* = s_c^*/s_c(T)$ to determine if the LCT provides insight into the correlations of relaxation data analyzed by Kivelson et al. [13], data that emphasize the relatively high temperature range where z^* first deviates from unity rather than the range where s_c becomes critically small.

Figure 7 presents z^* as a function of the reduced temperature variable $\delta T_A \equiv |T - T_A|/T_A$. The different curves of Fig. 7 refer to the F-F and F-S polymer classes and the same M as in Fig. 6. Evidently, the calculated z^* grows much faster with δT_A for the F-S polymer class and increases somewhat with M within each polymer class. These trends, taken in conjunction with the AG model, again translate into the prediction that polymer chains with bulky stiff side groups (F-S class) have a stronger dependence of the relaxation time on temperature (i.e., they are more fragile) than flexible chains with

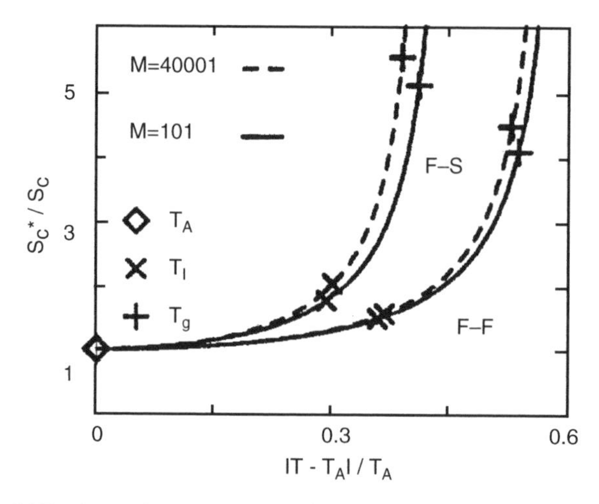

Figure 7. LCT estimates for the size $z^* = s_c^*/s_c$ of the cooperative rearranging regions (CRRs) in constant pressure ($P = 1$ atm; 0.101325 MPa) glass-forming polymer fluids as a function of the reduced temperature $\delta T_A \equiv |T - T_A|/T_A$ defined relative to the Arrhenius temperature T_A. Different curves refer to the F-F and F-S polymers with low and high molar mass ($M = 101$ and $M = 40001$). The characteristic temperatures of glass formation—T_g, T_I, and T_A—are designated in the figure by $+$, $\times$, and $\diamond$, respectively, for better visualization of $z^*(T)$ values. The dependence of z^* on δT_A is quantitatively parabolic. (Used with permission from J. Dudowicz, K. F. Freed, and J. F. Douglas, *Journal of Chemical Physics* **123**, 111102 (2005). Copyright © 2005, American Institute of Physics.)

flexible side groups (F-F class). This general finding is further amplified and quantified next.

C. Identification of Characteristic Temperatures

Since the magnitude of z^* directly affects structural relaxation times τ, the combination of our method for determining z^* with the AG model provides a means for understanding regularities in the order of magnitude of τ at T_I, T_A, and T_g. Table I summarizes the computed z^* at the three characteristic temperatures T_A, T_I, and T_g, which are designated in Fig. 7 by $\diamond$, $\times$, and $+$, respectively. At the crossover temperature T_I, for instance, the calculated value $z^* \approx 2$ for the F-S class of polymers accords well with the near "universal" value ($z^* \approx 2$) estimated by Ngai [15] for a variety of glass-forming liquids (e.g., toluene, orthoterphenyl, propylbenzene) near the crossover temperature T_{mc}^{exp} (sometimes denotes as T_B or T_c) between the low temperature VFTH regime near T_g and the high temperature regime where a different temperature dependence of τ applies [118]. Consequently, our definition of T_I exactly corresponds to this separation between high and low temperature regimes of glass formation, and we thus identify T_{mc}^{exp} with T_I of our entropy theory. Ngai [15] further notes that T_{mc}^{exp} closely coincides with the temperature T_β at which α and β relaxation times bifurcate, exhibiting a coexistence of noncollective and significant collective motions ($z^* = 2$). At least five different characteristic temperatures (all of comparable magnitude and derived from a variety of measurements) have been identified with the crossover between the high and low temperature glass regimes [119].

The apparently parabolic dependence of z^* on $\delta T_A = (T_A - T)/T_A$ in Fig. 7 is actually quantitative. For both classes of polymers, z^* can be described over a broad temperature range below T_A by the simple expression

$$z^* - 1 = C_s \left[\frac{|T - T_A|}{T_A} \right]^\alpha, \quad \alpha = 2, \quad T_I < T_A - 100\,\mathrm{K} < T < T_A \tag{36}$$

where the parameter C_s is summarized in Table I and the accuracy of the fit over the stated temperature range is better than 0.1%. A somewhat larger exponent α is obtained if the fitting is extended to temperatures lower than $T_A - 100$ K, but this extended range compromises the simplicity and generality of Eq. (36). Insertion of Eq. (36) into the AG equation for τ (and taking $\Delta\mu$ as constant) clearly yields a form different from the VFTH equation. In Section X, we advocate the use of C_s of Eq. (36) as a measure of glass fragility in the high temperature regime of glass formation.

Remarkably, the LCT estimates of the reduced activation energy z^* are compatible with the recent experimental estimates of $\mathcal{E}/\Delta\mu$ for diverse fluids by Kivelson et al. [13]. The particular exponent 8/3 reported by Kivelson et al. [13] for α in their analogue of Eq. (36) has been motivated by a "frustrated-limited cluster model" of glass formation, while the experimental exponent is somewhat uncertain, that is, the stated uncertainty is ± 0.3 about a "best" value near 8/3. We conclude that the AG model of glass formation seems to provide a promising description of structural relaxation times in glass-forming liquids for the entire temperature range from T_A, where the fluid first becomes "complex," down to T_0 (where it becomes an amorphous solid in the entropy theory). We further comment on the physical significance of the Arrhenius temperature T_A before discussing additional features of the configurational entropy s_c.

The onset of glass formation in a polymer melt is associated with the development of orientational correlations that arise from chain stiffness. At the temperature T_A, there is a balance between the energetic cost of chain bending and the increased chain entropy, and below this temperature orientational correlations are appreciable while the melt still remains a fluid. Such a compensation temperature has been anticipated based on a field theoretic description of semiflexible polymers by Bascle et al. [120]. The temperature T_A is important for describing liquid dynamics since the orientational correlations (and dynamic fluid heterogeneities associated with these correlations) should alter the polymer dynamics for $T < T_A$ from the behavior at higher temperatures where orientational correlations are absent [120]. Thus, the temperature T_A signals the *onset* of the correlated motions characteristic of glass formation. (The observation of complex fluid behavior, however, is possible only when the fluid does not crystallize for $T \leq T_A$.) Our designation of this onset temperature by T_A reflects the interpretation of this temperature as the point at which an Arrhenius dependence of structural relaxation times emerges as a reasonable description for higher temperatures. (Rough estimates of T_A for a wide variety of polymer fluids are tabulated by Aharony [121].) At lower temperatures, structural relaxation becomes nonexponential, relaxation times become extremely large, and large scale mobility fluctuations appear in the fluid (i.e., the fluid becomes "dynamically heterogeneous," see Section XII). Recent work [122, 123] provides support for this interpretation of T_A.

It is apparent from Fig. 6 and Table I that T_A tends to be higher in (fragile) F-S polymer fluids than in the (stronger) F-F polymers. Specifically, the computed values of T_A correspond reasonably well with the estimates from MC simulations and measurements suggesting that T_A lies in a broad range from 450 to 750 K for polymer fluids [108, 121, 124, 125]. The observation of collective fluid dynamics below T_A implies that the influence of incipient glass formation generally dominates transport processes in polymer fluids since organic polymers tend to become unstable due to thermal degradation at temperatures comparable to or somewhat greater than T_A. The high temperature regime is more accessible for highly flexible polymers, such as PDMS and many polyolefins, for which T_K and other characteristic temperatures are much lower.

While the excess entropy $S_{exc}^{(mol)}$ does not have a maximum at T_A, there are other thermodynamic signatures that seem to occur in the vicinity of this characteristic temperature. For example, Grimsditch and Rivier [126] have observed maxima in the ratio of the specific heats at constant pressure and volume C_P/C_V at temperatures of $1.9T_g$ and $1.7T_g$ for $ZnCl_2$ and glycerol, respectively. Moreover, the specific heat C_P normally exhibits a *minimum* for a comparable temperature range [13, 127, 128]. The temperatures at which these thermodynamic features occur are compatible with the LCT estimations of T_A/T_g in Table I, so that it seems likely that they are related to T_A. More quantitative studies of this correspondence, however, require a reliable calculation of the vibrational and other nonconfigurational contributions to the fluid entropy, which is beyond the scope of the present chapter. We recommend that T_A can be experimentally estimated by determining the entropy density $S_{exc}^{(mol)}/V^{(mol)}$ and by locating a maximum in this quantity as in our LCT definition of T_A.

D. *M*-Dependence of Characteristic Temperatures

Our theory of polymer melt glasses distinguishes four characteristic temperatures of glass formation that are evaluated for a given pressure from the configurational entropy $s_c(T)$ or the specific volume $v(T)$. Specifically, these four characteristic temperatures for glass formation are the ideal glass transition temperature T_0, where s_c extrapolates to zero, the kinetic glass transition temperature T_g, the crossover temperature T_I between the high and low temperature regimes of glass formation, and the Arrhenius temperature T_A. These characteristic temperatures are determined, respectively, from the following conditions:

$$s_c(P, T = T_0) = 0 \tag{37}$$

$$\frac{v(P, T = T_g) - v(P, T = T_0)}{v(P, T = T_g)} = \begin{cases} 0.027 \text{ F-S and S-F classes} \\ 0.016 \text{ F-F class} \end{cases} \tag{38}$$

$$\left.\frac{\partial^2 [s_c(P, T)T]}{\partial T^2}\right|_{P, T = T_I} = 0 \tag{39}$$

$$\left.\frac{\partial s_c(P, T)}{\partial T}\right|_{P, T = T_A} = 0 \tag{40}$$

The constraint in Eq. (38) that enables the direct computation of T_g is obtained by the extension of the Lindemann criterion to the "softening transformation" in glass-forming liquids [42, 56, 129, 130], and the details of this relation are explained in Section VI. Within the schematic model for glass formation (with specified ϵ, E_b, E_s, and monomer structure), all calculated thermodynamic properties depend only on temperature T, on pressure P, and on molar mass M_{mol} (which is proportional to the number M of united atom groups in single chains). The present section summarizes the calculations for T_0, T_g, T_I, and T_A as functions of M for a constant pressure of $P = 0.101325$ MPa (1 atm).

Figures 8a and 8b display the polymer mass dependence of T_A, T_I, T_g, and T_0 for the F-F and F-S classes of polymers, respectively. All four characteristic temperatures exhibit the same physical trend, that is, growing with increasing M and saturating to constants in the high molar mass limit $M \to \infty$. Comparison of Figs. 8a and 8b reveals that the variation of these characteristic temperatures with M is generally stronger for the F-S polymers. For instance, the slope of T_g versus $1/M$ is nearly a factor of 3 larger for the F-S class than the F-F class. All four characteristic temperatures T_α ($\alpha \equiv 0$, g, I, A) for the F-S polymers exceed their counterparts for the F-F polymers.

The molar mass dependence of the characteristic temperatures for glass formation is somewhat obscured by the large temperature scales used in Figs. 8a and 8b, but these variations with M are certainly observable and have many practical implications (affecting the stability of films, mechanical properties of polymeric materials, the viscosity of polymer fluids under processing conditions, etc.) Figures 8a and 8b are therefore replotted in semilog format to visually amplify the M dependence of $\{T_\alpha\}$, following the procedure

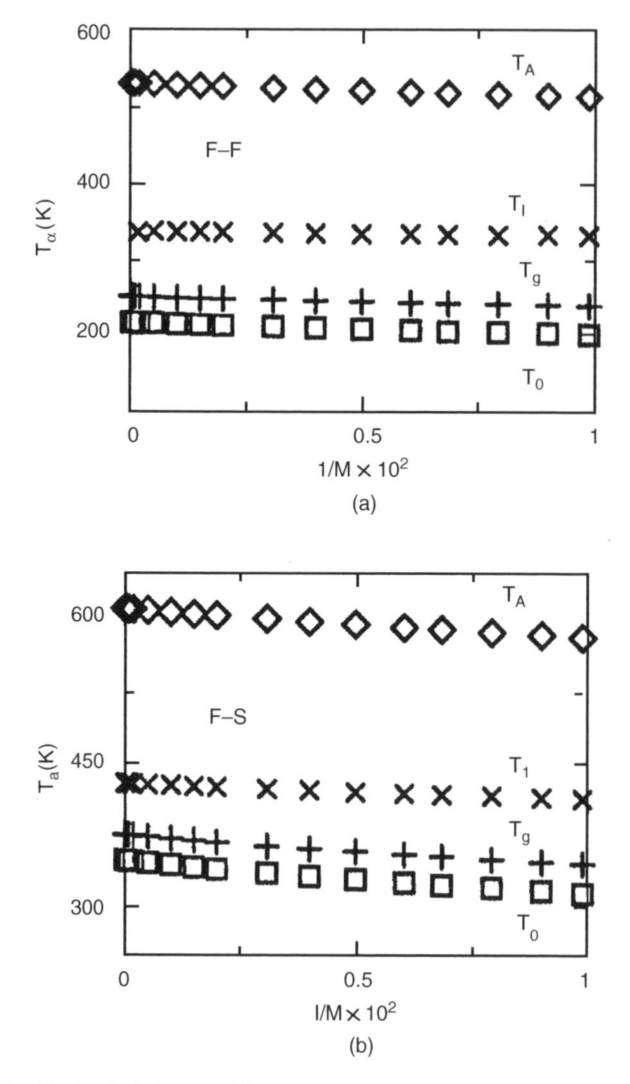

Figure 8. (a) The ideal glass transition temperature T_0, the glass transition temperature T_g, the inflection temperature T_I, and the Arrhenius temperature T_A, calculated from Eqs. (37), (38), (39), and (40), respectively, as a function of the inverse number $1/M$ of united atom groups in individual chains for constant pressure ($P = 1$ atm; 0.101325 MPa) F-F model monodisperse polymer fluids. (b) Same as (a) but for F-S polymer fluids.

employed in presenting experimental data [131] for T_g. The characteristic temperatures in Figs. 9a and 9b are derived from Figs. 8a and 8b, respectively, by normalizing T_α with the corresponding high molar mass limits $T_\alpha(M \rightarrow \infty) \equiv T_\alpha^\infty$, while the variable M is divided by the "crossover value"

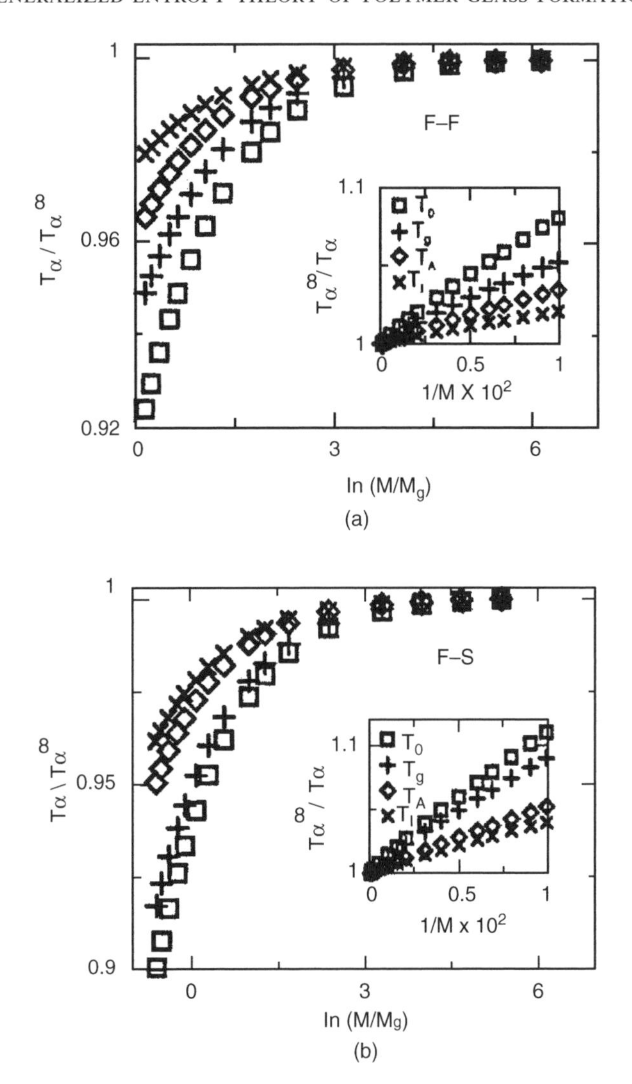

Figure 9. (a) Figure 8a replotted in a semilog format to amplify visually the M dependence of the characteristic temperatures $\{T_\alpha\}$. The temperatures T_0, T_g, T_I, and T_A are normalized by their corresponding high molar mass limits $T_\alpha(M \to \infty) \equiv T_\alpha^\infty$, while the variable M is divided by the value M_g at which the ratio T_g/T_g^∞ nearly saturates, that is, where $T_g(M = M_g)/T_g^\infty = 0.95$. The saturation value M_g is 86 for F-F polymers. The inset emphasizes a universal linear scaling for the inverse ratios T_α^∞/T_α with $1/M$. (b) Same as (a) but for F-S polymer fluids ($M_g = 184$). Parts (a) and (b) used with permission from J. Dudowicz, K. F. Freed, and J. F. Douglas, *Journal of Physical Chemistry B* **109**, 21285 (2005). Copyright © 2005 American Chemical Society.)

M_g at which the ratio T_g/T_g^∞ nearly saturates, that is, where $T_g(M = M_g)/T_g^\infty = 0.95$. A stronger mass dependence of T_g for the F-S polymers implies a larger value of M_g. It seems likely that relative values of M_g should provide an effective measure of the fragility of polymer fluids, but further tests are necessary to confirm this hypothesis.

The magnitude of the molar mass dependence differs for the characteristic temperatures T_α. Figures 9a and 9b show that T_0 and T_I have the strongest and the weakest dependence on M, respectively. This latter feature is also noticable from the insets to Figs. 9a and 9b, which display the inverse ratios T_α^∞/T_α as functions of $1/M$. The ratios T_α^∞/T_α *universally* saturate to unity in a linear fashion with $1/M$, but their slopes vary between the different T_α. (The linearity of the scaling is less accurate for the inverse ratio T_α/T_α^∞.) The insets also confirm the earlier finding that the strength of the M dependence of the T_α for the F-S polymers generally exceeds that for the F-F chains, in accordance with experimental observations [79, 132].

V. FRAGILITY OF GLASS-FORMING LIQUIDS

This section discusses the implications of the temperature dependence of the configurational entropy s_c (displayed in Figs. 6 and 7) on calculations of the fluid fragility that emerge from the AG model. The mutual consistency of the AG equation (33) and the VFTH equation (34) requires a linear scaling between the product $s_c(T)T$ and the reduced temperature δT. Figure 10 illustrates the relation between $s_c(T)T$ and δT over a broad temperature range (up to at least 100 K above T_0) for both the F-F and F-S classes of polymers and for small and large molar masses M_{mol}. Dimensionless variables are introduced by normalizing the configurational energy s_cT by the thermal energy k_BT_0 at the ideal glass transition temperature. It is apparent that s_cT varies approximately linearly with δT over the temperature range indicated in Fig. 10 for both the F-F and F-S polymers and that the slopes are quite *insensitive* to polymer mass. However, the slopes develop an appreciable dependence on M if s_cT is alternatively normalized [92] by the van der Waals energy ϵ, indicating a strong correlation between rate of change of $s_c(T)T$ with temperature ("thermodynamic fragility") and the magnitude of T_0. A similar reduction in the M dependence to that in Fig. 10 is obtained by normalizing s_cT by T_g. These observations similarly suggest a strong correlation between fragility and T_g, as recently noted by Novikov and Sokolov [133].

Because the AG equation (33) for τ reduces exactly to the VFTH equation (34) over the temperature range in which s_cT is proportional to δT, the correspondence between these expressions for τ uniquely establishes a relation between the kinetic fragility parameter $D \equiv 1/K_s$ and thermodynamic fragility $s_c(T)T/\delta T$. Specifically, in the temperature regime near the glass transition

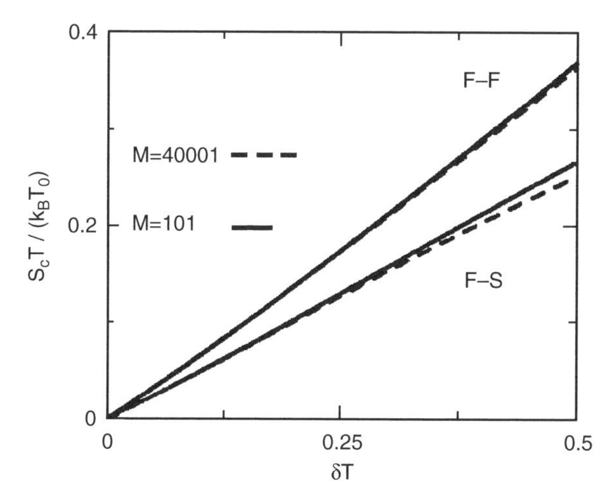

Figure 10. LCT configurational entropy s_cT as a function of the reduced temperature $\delta T \equiv (T - T_0)/T_0$ for low and high molar mass F-F and F-S polymer fluids at constant pressure of $P = 1$ atm (0.101325 MPa). The product s_cT is normalized by the thermal energy k_BT_0 at the ideal glass transition temperature T_0. (Used with permission from J. Dudowicz, K. F. Freed, and J. F. Douglas, *Journal of Physical Chemistry B* **109**, 21350 (2005). Copyright © 2005 American Chemical Society.)

where $s_c(T)T \propto \delta T$, the fragility in the low temperature regime of glass formation equals

$$K_s = (s_cT/\delta T)/(\Delta\mu s_c^*/k_B), \quad T_0 < T < T_I \tag{41}$$

The evaluation of the fragility parameter K_s from the definition in Eq. (41) is evidently complicated by the necessity of determining $\Delta\mu$ (which generally depends on molar mass [84]) for polymer fluids. (A definition of fragility that is free of this complication is discussed in Appendix A.) While the LCT does not prescribe a recipe for calculating energetic parameters such as $\Delta\mu$, observations based on experimental and simulation data suggest a means for estimating $\Delta\mu$. For example, simulations of both binary Lennard-Jones mixtures (Kob–Anderson model) [122] and simple models of Lennard-Jones particle chains [124] indicate that $\Delta\mu/k_B \approx 6T_{mc}^{exp}$. A large body of data for the viscosity of glass-forming ionic melts also supports this approximation, although the uncertainties are more difficult to determine in this instance since only a rough correlation with T_{mc}^{exp} is specifically indicated [134]. While the theoretical interpretation of the phenomenological temperature T_{mc}^{exp} is uncertain [123], it does have a well defined physical significance as a crossover temperature [9, 119] separating the high and low temperature regimes of glass formation where τ exhibits qualitatively different (and non-Arrhenius) temperature dependence in each regime. The crossover temperature $T_{\rm I}$ of the entropy theory is clearly a close counterpart of $T_{\rm mc}^{\rm exp}$, and the direct comparison of LCT computations for $T_{\rm I}/T_{\rm g}$ (see Table I) with literature estimates [91, 102] of the ratio $T_{\rm mc}^{\rm exp}/T_{\rm g}$ supports the identification of $T_{\rm I}$ with $T_{\rm mc}^{\rm exp}$. Specifically, $T_{\rm I}/T_{\rm g}$ equals 1.15 and 1.20 for high and low molar mass F-S chains, respectively, while $T_{\rm mc}^{\rm exp}/T_{\rm g}$ is 1.14 for the high molar mass PS and 1.18 for the model fragile small molecule liquid, orthoterphenyl [102]. The larger values of $T_{\rm I}/T_{\rm g}$ predicted for the F-F chains (1.35–1.39) are qualitatively consistent with the ratios of $T_{\rm mc}^{\rm exp}/T_{\rm g}$ for stronger fluids, but available data are largely restricted to ionic and hydrogen bonding fluids or to polymer melts whose glass formation is complicated by partial crystallization. Based on the identification of $T_{\rm mc}^{\rm exp}$ and $T_{\rm I}$, we estimate $\Delta\mu$ through the following simple approximation [135, 136]:

$$\Delta\mu/k_{\rm B} \approx 6T_{\rm I} \tag{42}$$

This relation enables evaluating the fragility parameter K_s as well as the structural relaxation times τ over the whole temperature range $T_0 < T < T_{\rm A}$. Because $T_{\rm I}$ depends on polymer microstructure and molar mass, $\Delta\mu$ likewise exhibits the same dependence. Computations of K_s within the entropy theory have not been possible before.

Our estimates of typical values for $\Delta\mu$ for both F-F and F-S high molar mass polymers ($\Delta\mu/k_{\rm B} \approx 2000$ K and 2600 K, respectively) are comparable in magnitude with $\Delta\mu$ obtained for high molar mass alkanes by Tabor [84] ($\Delta\mu/k_{\rm B} \approx 2700$ K). The interrelation between $\Delta\mu$ and $T_{\rm mc}^{\rm exp}$ has implications regarding the magnitude of the structural relaxation time τ at the crossover temperature $T_{\rm mc}^{\rm exp}$. Recent investigations [102, 137] indicate that τ at the crossover temperature $T_{\rm mc}^{\rm exp}$ is nearly "universal" for a large number of polymer glass formers, that is, $\tau(T_{\rm mc}^{\rm exp}) \sim \mathcal{O}(10^{-7\pm1}$ s). A similar regularity has been reported [15] for the enhancement of the apparent activation energy, $z^* = s_{\rm c}^*/s_{\rm c}(T)$, at $T_{\rm mc}^{\rm exp}$, namely, $z^*(T_{\rm mc}^{\rm exp}) \approx 2$. These observed regularities constrain the relation between $\Delta\mu$ and $T_{\rm I}$ in our theory. Inserting the above two values into the AG relation of Eq. (33) and taking the typical magnitude [102] for the high temperature limit of τ as $\tau_o \sim \mathcal{O}(10^{-13}$ s) lead to the conclusion that $\Delta\mu/k_{\rm B}$ should lie in the range

$$\Delta\mu/k_{\rm B} \approx (7 \pm 1)T_{\rm mc}^{\rm exp} \tag{43}$$

which is internally consistent with the empirical relation in Eq. (42) adopted previously.

Figures 11 and 12 analyze the variation of the fragility K_s with polymer class and with molar mass, respectively. The fragility parameter K_s is the slope of the curves depicted in Fig. 11. (Figure 11 departs from Fig. 10 only by the use of a

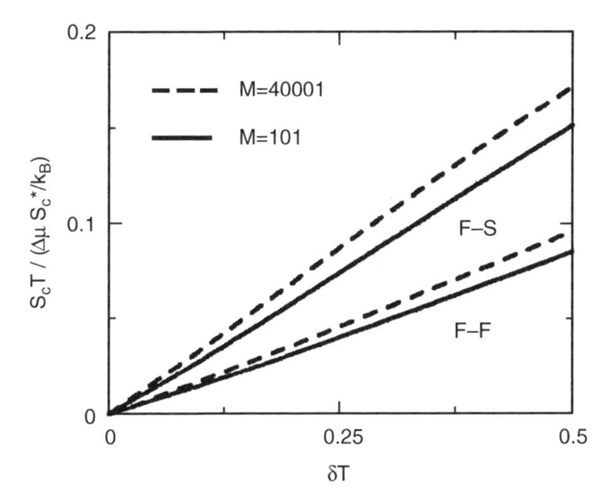

Figure 11. Same as in Fig. 10, but the configurational entropy $s_c T$ is normalized by the product of the critical entropy s_c^* and the activation energy $\Delta\mu$ (estimated from Eq. (42) and the computed crossover temperature T_I). According to Eq. (41), the slope of the resulting curves defines the fragility parameter K_s. (Used with permission from J. Dudowicz, K. F. Freed, and J. F. Douglas, *Journal of Physical Chemistry B* **109**, 21350 (2005). Copyright © 2005 American Chemical Society.)

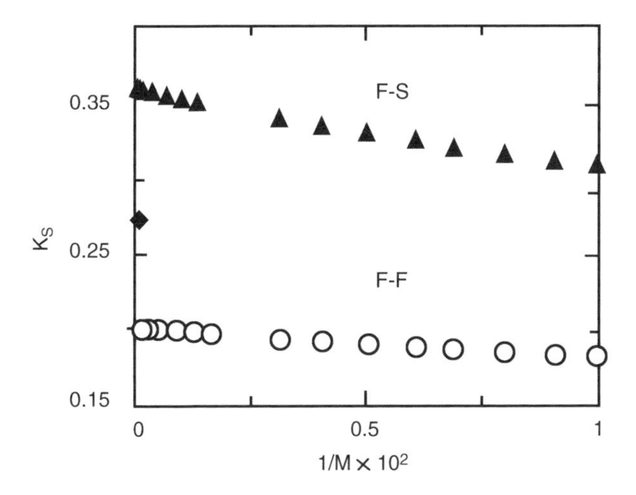

Figure 12. The fragility parameter K_s as a function of the inverse number $1/M$ of united atom groups in single chains for constant pressure ($P = 1$ atm; 0.101325 MPa) F-F and F-S polymer fluids. For a given M, the parameter K_s is determined as the slope of $s_c T/(\Delta\mu\, s_c^*/k_B)$ versus $\delta T \equiv (T - T_0)/T_0$ over the temperature range between T_g and T_I. A single data point denoted by ◆ refers to high molar mass F-S polymer fluid at a pressure of $P = 240$ atm (24.3 MPa). (Used with permission from J. Dudowicz, K. F. Freed, and J. F. Douglas, *Journal of Physical Chemistry B* **109**, 21350 (2005). Copyright © 2005 American Chemical Society.)

different normalizing factor for s_cT as prescribed by the by AG model and Eq. (41).) The slope defining K_s in Fig. 11 is definitely larger for the F-S polymer class than for the F-F class and depends somewhat on the polymer molar mass. This M dependence is quantified in Fig. 12, which shows that K_s first grows with M and then saturates for large M. (The high molar mass limit of K_s is summarized in Table I.) An increase in fragility of polystyrene with increasing M has been noted by Santangelo and Roland [79]. A similar behavior is obtained from the LCT for the variation of T_g and other characteristic temperatures of glass-forming fluids with M (see Section IV). Although recent measurements [138] indicate that the fragility of PIB decreases weakly with M, the observed dependence of fragility on M is indeed small, as would be expected for a F-F class polymer. This small deviation between the computed and observed M dependence may be explained by a number of secondary effects that are neglected in our schematic model of glass formation (e.g., monomer shape, tacticity, variability of interaction and bending energies with chemically different united atom groups).

Our entropy theory estimates of K_s in Figs. 11 and 12 compare quite reasonably with experimental data. For instance, the high molar mass limit of K_s for F-S polymers of $K_s^{\infty} = 0.36$ accords well with the value of 0.35 extracted by us from the data of Plazek and O'Rourke [139] for PS, which is a typical F-S class polymer. Some variability in the calculated K_s for PS appears, however, when the evaluation is based on the data tabulation of Ngai and Plazek [140]. An average of $K_s = 0.42 \pm 0.1$ is determined from four different data sets [140] for the stress-relaxation shift factor ($a_{T,\eta}$) for high molar mass glassy PS (where the uncertainty reflects the range of the data rather than the uncertainty in measurement). The rather large disparity in K_s emerges from variations in methodology (e.g., the assumption of time–temperature superposition, temperature interval investigated, polydispersity, tacticity, impurities). Similar comparisons of our estimates of K_s for F-F polymers ($K_s^{\infty} = 0.20$) are not straightforward because reliable data for K_s are indeed sparse due to the tendency of many F-F polymers to crystallize. Partial crystallization (not described by the LCT) renders both thermodynamic and transport properties highly sensitive to the cooling history and to other process variables, and values of K_s as large as 1 are sometimes found for systems that crystallize [141]. Literature data for glass transition temperatures (or VFTH parameters) are notoriously disparate and controversial for simple polymer fluids, such as polyethylene or polypropylene [140]. Polyisobutylene (PIB) is a well known "strong" polymer fluid that does not crystallize, and experimental estimates [140, 141] of K_s for PIB are normally much smaller than for PS, typically in the broad range 0.06–0.13.

The *relative values* of the characteristic temperatures are often suggested to provide insights into fluid fragility [102, 142, 143]. The ratios T_A/T_0 and T_A/T_g are measures of the breadth of the glass-formation process and thus of fragility.

Table I indicates that T_A/T_0 in the limit of high polymer molar mass is approximately 1.8 for the F-S class and 2.5 for the F-F class polymers, so that the glass transition temperature range is narrower on a relative basis for the more fragile F-S fluids. This trend seems to be consistent with tabulated values [13, 63] of T_A and T_0 (or its dynamical counterpart T_∞). For example, T_A/T_0 for the fragile glass-forming fluid (orthoterphenyl) is about 1.7, while this ratio equals about 2.5 and 2.7 for glycerol and *n*-propanol, respectively, which are known to be moderately strong liquids. Thus, T_A/T_0 appears to be a promising measure of fragility. The LCT estimates of T_A/T_0 should be useful in determining T_A for polymers when thermal degradation and other effects complicate the direct measurement of T_A.

The second ratio T_A/T_g equals 1.7 and 2.1 for the F-S and F-F classes, respectively, so that T_A/T_g is also larger for the stronger fluid. Curiously, this ratio nearly coincides with the common rule of thumb for T_m/T_g, where T_m is the melting temperature. (The rule applies to small molecule fluids [144], inorganic substances [145], and to numerous macromolecules [146] where T_m/T_g is claimed to equal 1.5 and 2 for symmetric and unsymmetric polymers, respectively.) This result supports the longstanding observation that T_m appears close to the temperature T_A where deviations from an Arrhenius temperature dependence start to occur [12]. Kivelson et al. [13] provide further evidence in favor of this correspondence. While it is unclear whether a fundamental connection exists between the maximum of the configurational entropy s_c at T_A and the tendency toward crystallization, this striking correlation is provocative. However, this question lies beyond the scope of the current investigation.

Figure 13 presents ratios of characteristic temperatures for the F-F and F-S polymers as slowly varying with $1/M$. The ratio T_I/T_g ranges from about 1.2 for the relatively fragile F-S polymer class to about 1.4 for the strong F-F polymer class (consistent with experimental observations of fragility [102, 131, 142, 143]), while T_A/T_0 ranges from 1.8 to 2.5 in Fig. 13 and is compatible with other literature estimates (mentioned before) and with the general trend that this ratio is larger for stronger glass-forming liquids [147]. Recent MD simulations by Riggleman et al. [J. Chem. Phys. **126**, 234903 (2007)] for model fragile and strong polymer fluids yield ratios of characteristic temperatures that agree rather well with those in Table I.

Another popular definition of fragility is likewise represented in terms of T_g and a crossover temperature, separating the high and low temperature regimes of glass formation (our T_I). This other definition has been used in recent attempts at constructing a universal reduced variable description for the temperature dependence of the viscosity [142] and the structural relaxation time [11] of glass-forming polymer liquids. The fragility parameter ψ in this context is defined as [11, 142]

$$\psi \equiv T_I/(T_I - T_g) \tag{44}$$

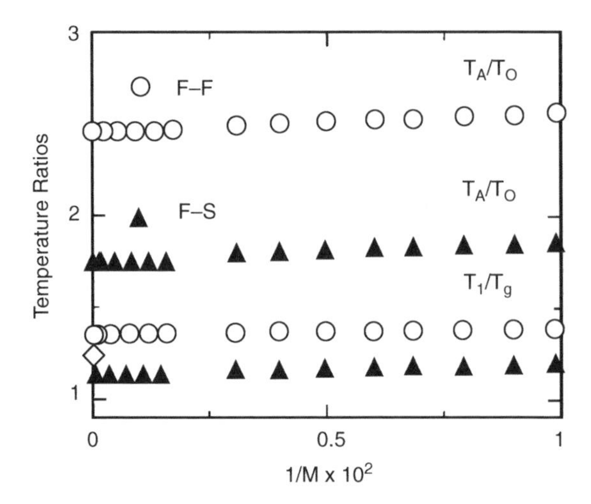

Figure 13. Ratios T_A/T_0 and T_I/T_g of the characteristic temperatures from Figs. 8a and 8b as a function of the inverse number $1/M$ of united atom groups in individual chains for constant pressure ($P = 1$ atm; 0.101325 MPa) F-F (open symbols) and F-S (filled symbols) polymer fluids. The single data point denoted by $\diamond$ refers to high molar mass F-S polymer fluid at a pressure of $P = 240$ atm (24.3 MPa).

where the crossover temperature is equated with the entropy theory temperature T_I, as explained earlier. To establish contact with experimental studies examining ψ, LCT calculations of ψ are illustrated in Fig. 14 as a function of $1/M$. The parameter ψ increases with M, saturates to a finite value

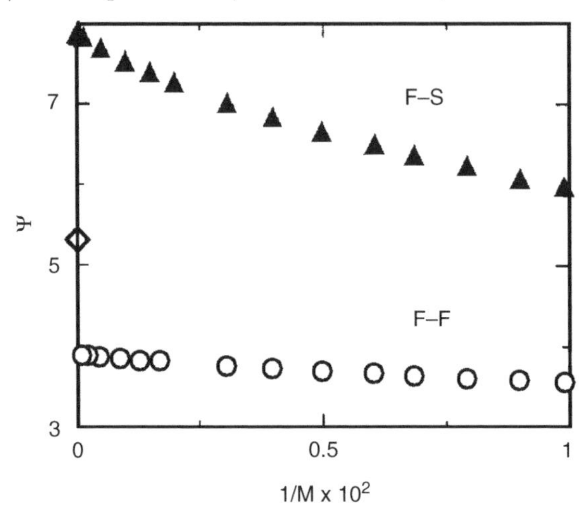

Figure 14. The fragility parameter $\psi = T_I/(T_I - T_g)$ calculated from the LCT as a function of the inverse number $1/M$ of united atom groups in individual chains for constant pressure ($P = 1$ atm; 0.101325 MPa) F-F and F-S polymer fluids. The single data point denoted by $\diamond$ refers to high molar mass F-S polymer fluid at a pressure of $P = 240$ atm (24.3 MPa).

at high M, and is larger for F-S polymers than for F-F chains, reflecting the greater fragility of the F-S polymers. Limiting values of ψ at high and low M are indicated in Table I. The range ($\psi \approx 5.5$ to 8) for ψ in Fig. 14 is reasonably consistent with reported literature data ($\psi \approx 7$ to 10) for a variety of polymer fluids [142]. Rössler et al. [142] demonstrate that ψ correlates strongly with the "steepness parameter" m (or dynamic fragility) determined from viscosity data near the glass transition,

$$m \equiv \left. \frac{\partial \ln[\eta(T)]}{\partial(T_g/T)} \right|_{T=T_g} \tag{45}$$

Specifically, the general approximation $\psi \approx 0.079\,\mathrm{m}$ has been suggested [142] for a wide range of both polymer and small molecule liquids. This relation implies that high molar mass F-S polymers should have steepness index m of about $m \approx 100$, which is consistent in order of magnitude with estimates [142] for the representative F-S polymer, polystyrene (PS).

Glass formation is evidently a phenomenon that is not restricted to high molar mass polymers. Orthoterphenyl, for example, can be considered as akin to a single monomer with aromatic side groups, and, indeed, the entropy theory fragility parameter K_s for low molar mass F-S polymers ($K_s = 0.31$) is consistent with the experimental value $K_s = 0.29$ reported by Richert and Angell [49]. Comparisons of the current LCT predictions for low molar mass polymers with the literature values of K_s for small molecule glass formers should, however, be taken with some caution because the shortest polymer chains considered by us ($M = 100$) are still long relative to small molecules consisting of several united atom groups. (In principle, our theory can describe glass formation in small molecule fluids, but the mean field approximation inherent to the LCT becomes less accurate for small M.) Similarly, we can view n-propanol as a representative member of the F-F class of monomers, and the agreement between the literature [49] $K_s = 0.18$ and the corresponding theoretical $K_s = 0.18$ from Fig. 12 seems even better. The fragility of glycerol [91], another strong liquid with rather simple structure, is somewhat lower ($K_s = 0.05$), however. Smaller values of K_s are characteristic of numerous sugars and other hydrogen bonding fluids [105]. This trend is understandable from Eq. (41), which indicates that K_s varies inversely with $\Delta\mu$, which, in turn, depends on the cohesive energy density or the strength of van der Waals interactions.

VI. FREE VOLUME VERSUS CONFIGURATIONAL ENTROPY DESCRIPTIONS OF GLASS FORMATION

Endless discussion exists regarding whether a theory based on the configurational entropy s_c or the excess free volume δv provides the more correct description of glass formation. Thus, this section briefly analyzes the relation between these thermodynamic properties within a common LCT framework. While the previous sections emphasize a description of glass formation in terms of variations of the configurational entropy, complementary information about the nature of the glass transition can be obtained from the LCT by focusing instead on the excess free volume δv. Specifically, a kinetic instability (Lindemann) criterion [42, 56, 129, 130] is introduced below to determine the glass transition temperature T_g based on LCT calculations for the temperature dependence of the specific volume. Notably, this definition of the glass transition temperature asserts that T_g is not accompanied by *any overt thermodynamic signatures* [148] and is thus distinct from the ideal glass transition temperature T_0 where s_c extrapolates to zero [149].

The reduced specific volume δv relative to its value at T_0,

$$\delta v(T) = [v(T) - v(T = T_0)]/v(T) \tag{46}$$

is a well defined macroscopic fluid property [150, 151] that quantifies how much "free space" exists for atomic motion in the polymer material. This definitions of δv does not include the residual unoccupied space that is frozen in at lower temperatures than T_0 and that contributes rather little to molecular movement. Moreover, the vanishing $\delta v(T)$ at T_0 is also consistent with a molecular scale definition of excess free volume based on Debye–Waller factors [152].

Calculations of δv from the LCT are straightforward. Equation (46), in combination with the LCT equation of state, enables expressing δv in terms of the polymer volume fraction $\phi(T)$,

$$\delta v(T) = [\phi(T) - \phi(T = T_0)]/\phi(T) \tag{47}$$

where $\phi(T)$ is computed from the LCT as a function of temperature T and pressure P (for a specified lattice cell volume v_{cell}).

Figure 15 presents LCT calculations of $s_c T$ and δv for the same energetic parameters, M and P, and the same F-F and F-S classes of polymers as in Figs. 6, 7, 10, and 11. The temperature range in Fig. 15 corresponds roughly to the same interval $T_0 < T < T_I$ as in Fig. 11, and for consistency, $s_c T$ is normalized by $(\Delta\mu s_c^*/k_B)$ according to Eq. (41). As expected from the parallel successes of the free volume and entropy theories in describing τ of glass-forming liquids, an approximately proportional relation is found between $s_c T$ and δv in Fig. 15. However, this linearity occurs over a more limited range of δv for the F-F polymers. The linear variation of $s_c T$ with δT persists over a substantially larger range (see Fig. 11) because of the nonlinear temperature dependence of δv. A surprising feature of Fig. 15 is the weak variation of the slopes in this figure with M compared to those in Fig. 11.

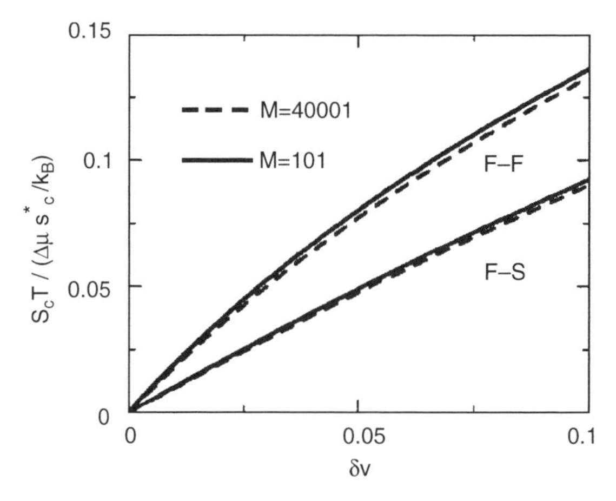

Figure 15. The LCT configurational entropy s_cT (normalized as in Fig. 11) as a function of the reduced specific volume $\delta v \equiv [v(T) - v(T = T_0)]/v(T)$, which is a measure of the excess free volume in the polymer system. Different curves refer to the F-F and F-S polymer fluids and to low and high molar mass polymer chains.

While the glass transition temperature T_g may be defined at this juncture as the temperature at which $\tau \approx 10^2$–10^3 s, an alternative, but equivalent, definition is employed here to emphasize the close relation between the entropy and free volume theories of glass formation. A knowledge of the temperature variation $\delta v(T)$ at constant pressure allows us to introduce a definition of the glass transition temperature T_g into the lattice model theory description of glass formation based on an extension of the Lindemann criterion, which provides a general estimate for the instability of the solid state to large scale thermally excited particle motions that lead to a sharp increase in molecular fluidity when the scale of these motions becomes critically large relative to the limiting average interparticle distance at low temperatures. Both theory and measurements support the extension of the Lindemann criterion to the "softening transformation" in glass-forming liquids [42, 56, 129, 130]. According to the Lindemann criterion, the root mean square amplitude of particle displacements at the glass softening or crystal melting temperature is on the order of 0.1 times the low temperature limiting interparticle distance $2R$, a condition often expressed as approximately 0.15 times $2R$. The particular value depends on type of ordering and on the nature of the intermolecular potential [42]. At one end of this range, the value ≈ 0.125 is characteristic of the melting of hard spheres [153, 154] (a reasonable model for molecules having simple symmetric geometrical structures), while values closer to 0.15, or even larger (0.17–0.185) are cited for particles having longer range interactions [129]. When applied to spherical particles of radius R in spherical cavities, this criterion implies that the ratio of the excess volume available for motion of the center of the particle and the particle's own limiting low temperature volume ranges from about $(0.25)^3 \approx 0.016$ to $(0.3)^3 = 0.027$. Correspondingly, we define the glass transition temperature T_g by the condition that the relative free volume $\delta v(T)$ achieves the high end of this range (i.e., $\delta v(T = T_g) = 0.027$) for the relatively fragile F-S class and an intermediate value $\delta v(T = T_g) = (0.25)^3 \approx 0.016$ for the relatively strong F-F polymers. This choice is also heuristically motivated by the observation that the excess free volume is generally significantly smaller for the F-F class polymer fluids. These definitions of T_g lead to the computed relaxation times $\tau(T = T_g) \sim \mathcal{O}(10^2–10^3\ \text{s})$ for both classes of fluids and thus are also consistent with the standard phenomenological definition [3, 133] of T_g. Of course, the Lindemann criterion only provides rough estimates of the experimental T_g (at which fluid properties abruptly change in cooling measurements due to the system becoming "stuck" and going out of equilibrium) because the experimental T_g is an inherently uncertain quantity that depends on cooling rate, sample history, and so on. Essentially, the same glass transition criterion as ours for fragile F-S polymers ($\delta v(T = T_g) \equiv 0.025$) has been suggested empirically long ago by Ferry and co-workers [75]. Finding the precise prescription for defining the glass transition temperature T_g completes our effort to evaluate and interrelate the four characteristic temperatures (T_0, T_g, T_I, T_A) associated with glass formation in polymer fluids.

VII. ISOTHERMAL COMPRESSIBILITY, SPECIFIC VOLUME, SHEAR MODULUS, AND "JAMMING"

In addition to the configuration entropy, the LCT enables the evaluation of other basic thermodynamic properties of polymer fluids, such as the density ρ (or specific volume v), isothermal compressibility κ_T, and thermal expansion coefficient, as functions of pressure, temperature, and molar mass. An examination of the temperature dependence of these properties, especially ones sensitive to excess free volume, provides additional perspectives on glass formation that are not apparent from a description based on s_c alone. Specifically, we demonstrate that both κ_T and v exhibit a stronger temperature and molar mass dependence for the F-S class than for the F-F class of polymers, so that variations of fragility with the rigidity of the side groups deduced by analyzing s_c are also reflected in these properties. Moreover, the computed isothermal compressibility κ_T decreases relatively sharply as the glass transition is approached from higher temperatures, and this decrease is accompanied by a corresponding growth in elastic constants, such as the high frequency shear modulus [155–157] G_∞. The trends in the temperature dependence of κ_T and G_∞ ultimately lead to a strong suppression of thermal molecular motions at T_g and to the formation of a "jammed" nonequilibrium state. These additional thermodynamic properties

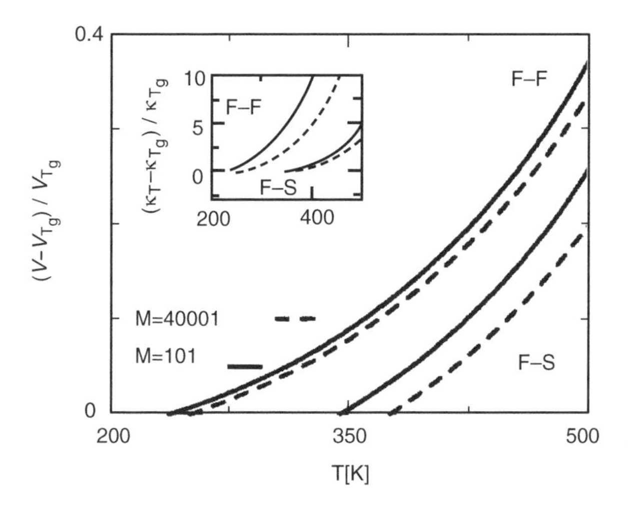

Figure 16. Reduced specific volume $(v - v_{T_g})/v_{T_g}$ and reduced isothermal compressibility $(\kappa_T - \kappa_{T_g})/\kappa_{T_g}$, defined relative to their values at the glass transition temperature T_g and calculated from the LCT as functions of temperature T for constant pressure ($P = 1$ atm; 0.101325 MPa) polymer fluids. Different curves refer to the F-F and F-S classes of polymers and to different numbers M of united atom groups in single chains. (Used with permission from J. Dudowicz, K. F. Freed, and J. F. Douglas, *Journal of Physical Chemistry B* **109**, 21285 (2005). Copyright © 2005 American Chemical Society.)

thus provide complementary information to s_c that aids in understanding the physical context and limitations of the entropy theory.

Figure 16 illustrates the temperature variation of the specific volume and isothermal compressibility for F-F and F-S polymers and for the same representative molar masses ($M = 101$ and $M = 40001$) as in Figs. 6, 7, 10, and 11. Since measurements of both v and κ_T are possible only above T_g, we designate T_g as the reference point in Fig. 16 and introduce the dimensionless variables $\delta v_g = [v(T) - v(T = T_g)]/v(T = T_g)$ and $\delta\kappa_{T_g} = (\kappa_T - \kappa_{T_g})/\kappa_{T_g}$. Similar reduced variables are used in Fig. 17 to display the molar mass (M) dependence of $v(T = T_g)$ and κ_T at T_g. Although the temperature dependence of v and κ_T in Fig. 16 could be regarded as very roughly linear, substantial curvature is apparent, especially over a large temperature range and for the F-F class of polymers (as observed experimentally [150, 158] for polyethylene (PE) and polydimethylsiloxane (PDMS)). The fairly linear variation of $v(T = T_g)$ and κ_{T_g} with $1/M$ in Fig. 17 is general for numerous other thermodynamic properties, regardless of polymer type (thus applying to both the F-S and F-F polymers and to other generalizations of these classes). Moreover, the characteristic temperatures $\{T_\alpha\}$ also share this fairly linear M dependence, as described in Section IV. Figure 17 also reveals a stronger M dependence of

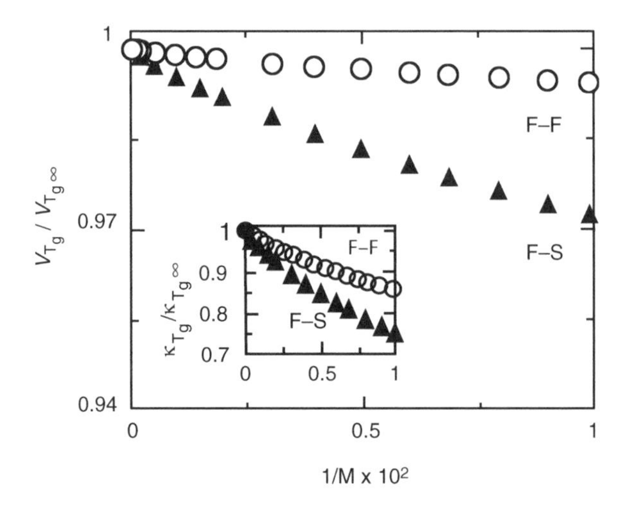

Figure 17. Specific volume v_{T_g} and isothermal compressibility κ_{T_g} (at the glass transition temperature T_g) calculated from the LCT as a function of the inverse number $1/M$ of united atom groups in single chains for constant pressure ($P = 1$ atm; 0.101325 MPa) F-F and F-S polymer fluids. Both quantities are normalized by the corresponding high molar mass limits (i.e., by $v_{T_g}^{\infty}$ or $\kappa_{T_g}^{\infty}$).

both $v(T = T_g)$ and κ_{T_g} for the more fragile F-S polymers than for F-F chains, and the same pattern is exhibited by T_g and other characteristic temperatures. Another nontrivial behavior of glass-forming fluids that emerges from Fig. 17 is the increase of the specific volume $v(T = T_g)$ with M. This counterintuitive trend has been confirmed experimentally [159, 160] and arises mainly from the fact that T_g increases with M. Finally, a knowledge of v and κ_T enables the estimation of the interfacial tension γ, which is expressed as proportional to $(\kappa_T v)^{1/2}$ in a Landau theory calculation by Sanchez [161]. (This simple and useful relation has been verified for PDMS and numerous small molecule liquids [162].)

The variation of κ_T and $v(T)$ with T and M (in Figs. 16 and 17) are both consequences of the increasing excess free volume (i.e., $\phi_v \equiv 1 - \phi$) in the fluid when T and $1/M$ are increased. The excess free volume ϕ_v is found to be higher for F-S polymers with their rigid side groups than ϕ_v for F-F class polymers at the same T and P, implying that both thermodynamic and dynamic properties of F-S polymers are more susceptible to temperature changes. In other words, F-S class polymers are more fragile than F-F chains because the former have higher specific volumes and isothermal compressibilities at T_g (see Table I). LCT calculations for the influence of pressure on glass formation (see Section X) show that increasing pressure (which causes a decrease of ϕ_v) leads to decreased fragility, in accord with the trends in Figs. 16 and 17 and with the interpretation of fragility as being influenced by variations of ϕ_v with temperature.

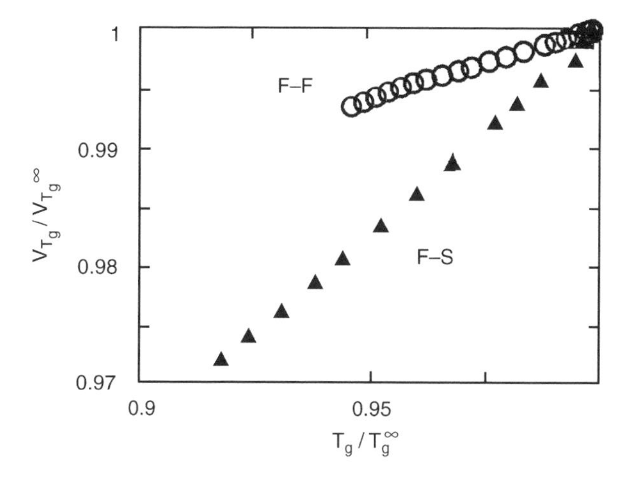

Figure 18. Variation of the specific volume v_{T_g} at the glass transition temperature T_g with the glass transition temperature T_g as calculated from the LCT for constant pressure ($P = 1$ atm; 0.101325 MPa) F-F and F-S polymer fluids. Both v_{T_g} and T_g are normalized by the corresponding high molar mass limits ($v_{T_g}^{\infty}$ or T_g^{∞}). (Used with permission from J. Dudowicz, K. F. Freed, and J. F. Douglas, *Journal of Physical Chemistry B* **109**, 21285 (2005). Copyright © 2005 American Chemical Society.)

One recognized and striking observation for glass-forming liquids provides a good test of the entropy theory, namely, the finding that the density ρ at T_g tends to *decrease* with increasing molar mass [159]. This trend is opposite to that observed at fixed temperatures much higher than T_g, where ρ increases monotonically with molar mass [131]. More specifically, measurements indicate [159] that the reciprocal of the density (specific volume) at T_g is itself nearly proportional to T_g, suggesting a remarkable correlation between these quantities. Figure 18 demonstrates that this empirical correlation between $v(T = T_g)$ and T_g emerges from the LCT calculations for both F-S and F-F polymers and that the slope is substantially larger for the more fragile F-S class of polymers. These results strongly imply that a higher T_g and a higher fragility can be associated with a higher free volume in the glassy state, which, in turn, arises due to the frustration in packing more complex shaped or extended molecules.

Experiments indicate that the smooth variations of thermodynamic properties (e.g., v, κ_T, and the specific heat at constant pressure C_P) with temperature are interrupted by the kinetic process of glass formation, leading to cooling rate dependent "kinks" in these properties as a function of temperature. In our view, these kinks cannot be described by an equilibrium statistical mechanical theory, but rather are a challenge for a *nonequilibrium* theory of glass formation. Nonetheless, some insight into the origin of these kinks and the qualitative meaning of the fluid becoming stuck can be gleaned from considering the magnitude of the isothermal compressibility κ_T as T_g is approached.

The high frequency shear modulus G_∞ is an equilibrium fluid property that is closely related [155] to κ_T since G_∞ describes the mean square amplitude of particle displacements about their quasiequilibrium positions in the dense fluid, while κ_T reflects the mean square amplitude of density fluctuations. The reciprocal of κ_T is normally termed the "bulk elasticity modulus" because it can be viewed as the fluid analogue of the bulk modulus of an elastic material [163]. The rough proportionality between the shear modulus and the bulk modulus implies the approximate scaling relation $G_\infty \sim 1/\kappa_T$ in the "glassy" regime $T_g < T < T_I$. (Indeed, computed values of $1/\kappa_T$ at T_g (see Table I) are on the order of 1 GPa, which is a typical order of magnitude for G_∞ and for $1/\kappa_T$ near the glass transition for both F-S (PS) and F-F (PP, PDMS) classes of polymers [157, 158, 164].) Cooling a liquid leads to a steady decrease in the fluid compressibility κ_T, as displayed in Fig. 16, and correspondingly to a rise in the high frequency shear modulus G_∞ [156, 163, 165]. The increase in the "stiffness" of glass-forming liquids ultimately becomes so large that it causes the molecular motions associated with thermal fluctuations to become "frozen" at the glass transition [163]. (This "inertial catastrophe" viewpoint of glass formation is briefly discussed by Starr et al. [152].) Because of this structural instability, the high frequency shear modulus G_∞ and the specific volume vary [156, 163, 165] more slowly with temperature below T_g, while κ_T decreases relatively sharply. These important changes in apparent thermodynamic properties are not captured by the thermodynamic theory since they arise from the extremely congested nature of the molecular motions and reflect the rate of cooling in the measurements. The decrease in κ_T in the vicinity of T_g is a signal that the fluid has entered a "jammed" nonequilibrium state with solid-like characteristics, corresponding to a kind of "death rattle" of the liquid state. These brief considerations of the implications of compressibility changes in cooled liquids provide additional insight into the nature of glass formation that are not directly describable by the entropy theory that focuses on the temperature variation of s_c. The above analysis also points to basic limitations of an equilibrium theory that must be respected in comparisons with experiment.

VIII. INFLUENCE OF SIDE GROUP SIZE ON GLASS FORMATION

Our discussion has so far been restricted to the schematic model of glass formation, which focuses on the relative flexibility of the chain backbone and side groups. The side groups in this schematic model are short linear chains (see Fig. 3b) with three united atom units, a structure inspired by many synthetic polymers in which the size of the side groups is on the order of a few carbon–carbon bonds. Within this idealized model of polymer glass formation, the computed T_g generally increases with either E_s or E_b, in accord with physical intuition and experimental observations indicating that greater chain rigidity leads to a higher T_g. However, the rate of this increase in T_g depends on the relative magnitude of E_b and E_s (i.e., on the polymer class) and the side group length n ($n = 3$ in the schematic model). In this section, we examine the dependence of T_g on n when all other characteristic parameters (ϵ, E_b, E_s, and M) of the model are held *constant*.

Because a description of the contrasting influences of rigidity in the side groups and the chain backbone requires a consideration of chains with at least a pair of bonds in the side groups, Fig. 19a presents LCT calculations of T_g as a function of the side group length n for polymer chains having fixed total molar mass ($M = 40001$ united atom groups) and $n \geq 2$. An increase of n in F-S polymers leads to a sharp rise in T_g in Fig. 19a. This trend of an increasing T_g with more extended, rigid side groups seems to be quite general in our calculations and is consistent with recent measurements [166, 167] for poly(2-vinyl naphthalene), a system with fairly rigid and extended side groups. Specifically, experiments reveal that T_g of high molar mass poly(2-vinyl naphthalene) is 50 K larger than the T_g of polystyrene ($T_g^{\infty} = 373\,\text{K}$) [140], which has a smaller phenyl side group. The calculated n dependence of T_g is significantly weaker for the F-F polymer class, where the backbone and side group rigidities are small and similar. In both F-F and F-S cases, T_g grows monotonically with n and approaches the T_g of the purely linear chain whose bending energy E_b is the same as E_s of the side groups of the structured monomer chains. These asymptotic large n limits are indicated in Fig. 19a by dashed lines. Inspection of Fig. 19a suggests that the appearance of additional chain ends as n is reduced (at fixed M) leads to a decrease of T_g, relative to the T_g of the asymptotic linear chain, in accord with the simple free volume arguments of Fox and Loshaek [150]. Specifically, the free chain ends comprise (at constant M) a smaller fraction of the total chain segments as n grows, and T_g correspondingly rises toward its large n asymptote. The increase in T_g with n is accompanied by a decrease of the excess free volume concentration ϕ_v with n for a given $T > T_g$. A similar decrease in ϕ_v and increase in T_g arise when ϵ is raised at constant n. In *real* polymer systems, on the other hand, the van der Waals energies ϵ_{ij} vary with the different chemical groups and other effects (e.g., microstructure, tacticity). Since these features are not included in the schematic model of polymer glasses, deviations may appear from the trends in Fig. 19a, especially for F-F class polymers where the computed T_g varies weakly with n and where competing contributions are comparable in magnitude.

The predicted growth of $T_g(n, M = \text{constant})$ with n is not universal for all polymer classes. If a flexible side group is attached to a relatively stiff backbone, T_g drops with increased n since the longer side groups "plasticize"

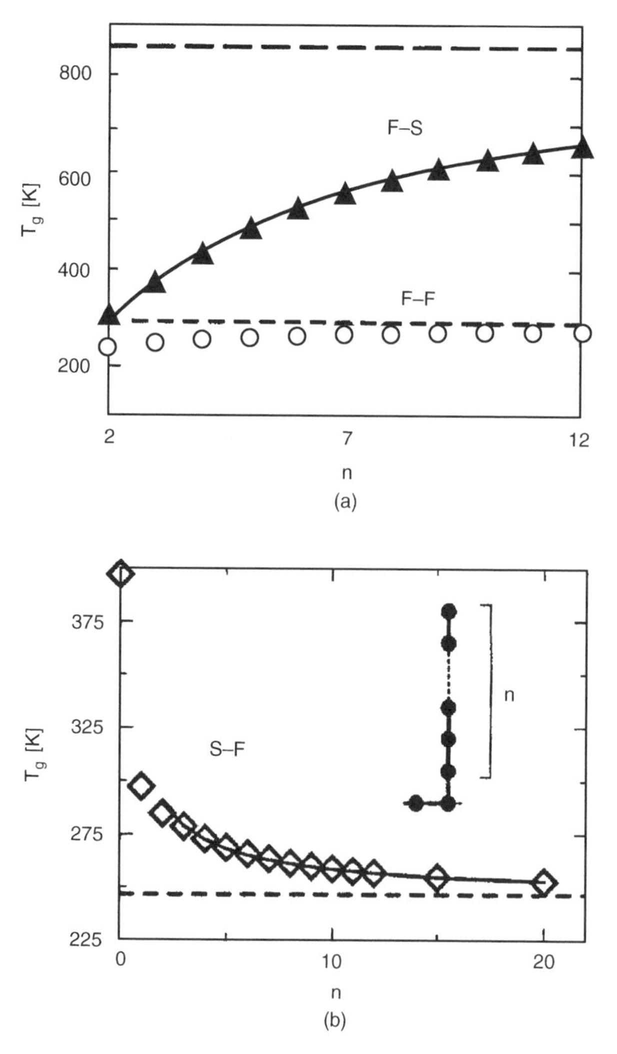

Figure 19. (a) Glass transition temperature T_g (symbols) as a function of the number n of united atom units in the side group as calculated from the LCT for constant pressure ($P = 1$ atm; 0.101325 MPa) F-F and F-S polymer fluids having fixed molar mass ($M = 40001$). Dashed lines indicate the glass transition temperatures of a melt composed of linear chains ($M = 40001$) whose bending energy E_b is the same as the bending energy E_s in F-F or F-S polymers. Solid line represents a fit to the data points for F-S polymers, using $T_g = an/(1 + bn)$ with $a = 215.8$ K and $b = 0.2422$. No simple fitting function has been found for $T_g(n)$ of F-F polymer fluids. (b) The same as (a) but for S-F polymer fluids. The solid line represents a fit to the data points (for $n = 3$–20), using $T_g = (a + bn)/(1 + n)$ with $a = 375.3$ K and $b = 246.9$ K. The dashed line denotes the glass transition temperature for a melt of linear chains whose bending energy E_b is the same as the bending energy E_s in S-F polymers. The inset depicts the monomer topology for the S-F, F-S, and F-F class polymers. (Parts (a) and (b) used with permission from J. Dudowicz, K. F. Freed, and J. F. Douglas, *Journal of Physical Chemistry B* **109**, 21285 (2005). Copyright © 2005 American Chemical Society.)

the stiff backbone. This trend for S-F polymers is illustrated in Fig. 19b, which exhibits a sharp decrease of $T_g(M = 40001)$ with n; T_g levels off for $n \sim \mathcal{O}(10)$; and finally T_g saturates to the T_g of the linear chain model for which $E_b = 200$ K and $M = 40001$. A very similar behavior is observed by Floudas and Štepánek [99] for an experimental counterpart of S-F polymers — a homologous series of poly(n-alkyl methacrylates). Increasing the side group length n (i.e., decreasing the fraction of free ends at constant M) again leads to a diminishing excess free volume concentration ϕ_v at a given $T > T_g$. However, this trend does not produce an increase of T_g for the S-F polymers, as the free volume arguments of Fox and Loshaek [150] would imply. Apparently, the manner in which free ends affect the glass transition temperature depends on the relative flexibility of the chain backbone and side groups. Nevertheless, when n becomes large in each of three classes of polymers (F-F, F-S, and S-F), the glass transition temperature generally approaches that of a melt of linear chains having a bending energy E_b equal to the E_s of the long chain side groups.

Our previous discussions of fragility indicate that glass fragility tends to increase with the excess free volume concentration ϕ_v (see Section VI). The general decrease in the calculated ϕ_v with growing n suggests that there is a common tendency toward reduced fragility as n becomes larger than 3. This suggestion is supported by experimental studies of poly(n-alkyl methacrylate) polymers, which exhibit stronger glass formation with increasing side group size [99, 168–170].

In summary, the side group length n significantly affects the magnitude of T_g. For fixed ϵ, stiffer and longer side groups lead to larger T_g, while the addition of more flexible side groups than the chain backbone can cause T_g to decrease. Our computations in conjunction with measurements of Floudas and Stepánek [99] and of Erwin and co-workers [166] indicate that controlling side group structure should provide a powerful means to regulate the T_g and the fragility of glass-forming polymers. It is reassuring that the entropy theory is able to predict these experimentally established trends in variations of T_g with molecular structure.

IX. TEMPERATURE DEPENDENCE OF STRUCTURAL RELAXATION TIMES

The dramatic slowing down of structural relaxation in glass-forming liquids lies at the heart of understanding the nature of glass formation. The entropy theory provides detailed predictions for the dependence of the structural relaxation time τ as a function of temperature, pressure, monomer structure, and molar mass. Importantly, the temperature range described by the theory is not restricted to the immediate vicinity of the glass transition temperature T_g, but comprises the entire, broad glass-formation regime ranging from the ideal glass transition temperature T_0 to the Arrhenius temperature T_A. The present section summarizes our theoretical predictions for the temperature variation of τ for F-F and F-S polymer fluids over this full temperature range at atmospheric pressure.

The LCT, in conjunction with the Adam–Gibbs equation (33) and the relation between the activation energy $\Delta\mu$ and crossover temperature T_I in Eq. (42), enables the computation of τ without any adjustable parameters beyond those inherent in the schematic model for glass formation (ϵ, E_b, E_s). Figures 20a and 20b illustrate the computed τ as a function of the reduced inverse temperature T_g/T for F-S and F-F polymers, respectively. The Arrhenius temperature T_A, crossover temperature T_I, and glass transition temperature T_g are denoted by the same symbols as in Figs. 6 and 7. The remaining characteristic temperature for glass formation, the VFTH temperature T_0 at which τ diverges, is indicated in the figures as a solid vertical line. Dashed lines refer to the high temperature regime $T > T_A$, where an Arrhenius relation $\tau = \exp(\beta\,\Delta\mu)$ is valid, and to very low temperatures $T < T_g$ where τ evidently becomes astronomical. The change in τ over the range $T_g < T < T_A$ in Figs. 20a and 20b is stupendous, about 15 and 14 orders of magnitude for F-S and F-F polymers, respectively. The predicted relaxation time at the onset of glass formation $\tau(T = T_A)$ is on the order of a few picoseconds (a few tens of molecular collision times) for both classes of polymers, while $\tau(T = T_I)$ at the crossover temperature T_I is several orders of magnitude larger, that is, $\mathcal{O}(10^{-8\pm1}\text{ s})$. The predictions in Figs. 20a and 20b are also reasonably consistent with recent measurements [102] of τ at the experimental mode coupling temperature T_{mc}^{exp} (the analogue of T_I) for a wide variety of glass-forming liquids where τ is found [102, 137] to be $\mathcal{O}(10^{-7\pm1}\text{ s})$. Figures 20a and 20b further indicate that relaxation times increase rapidly for temperatures below T_I, with $\tau(T = T_g)$ becoming as large as 10^3 s, a typical order of magnitude for glass-forming liquids [102]. The structural relaxation time τ of both the F-S and F-F polymer classes exhibits a similar temperature variation, although the steepness of the rise in τ at low temperatures is less pronounced for F-F polymers, reflecting the stronger nature of this class of fluids.

A further test of our identification of T_I with the empirical temperature T_{mc}^{exp} can be generated by fitting the constants A and γ of the mode coupling expression [19] for τ,

$$\tau = A(T - T_{mc}^{exp})^{-\gamma} \tag{48}$$

to the entropy theory calculations for τ over a temperature range above T_I. This functional form has been widely applied to reproduce experimental and simulation data for the structural relaxation time for glass-forming liquids over a limited temperature range above T_{mc}^{exp}. We find that Eq. (48) describes the entropy theory prediction for τ with an accuracy better than 0.15% over the temperature ranges 379–451 K and 463–520 K, with the *apparent exponent* γ

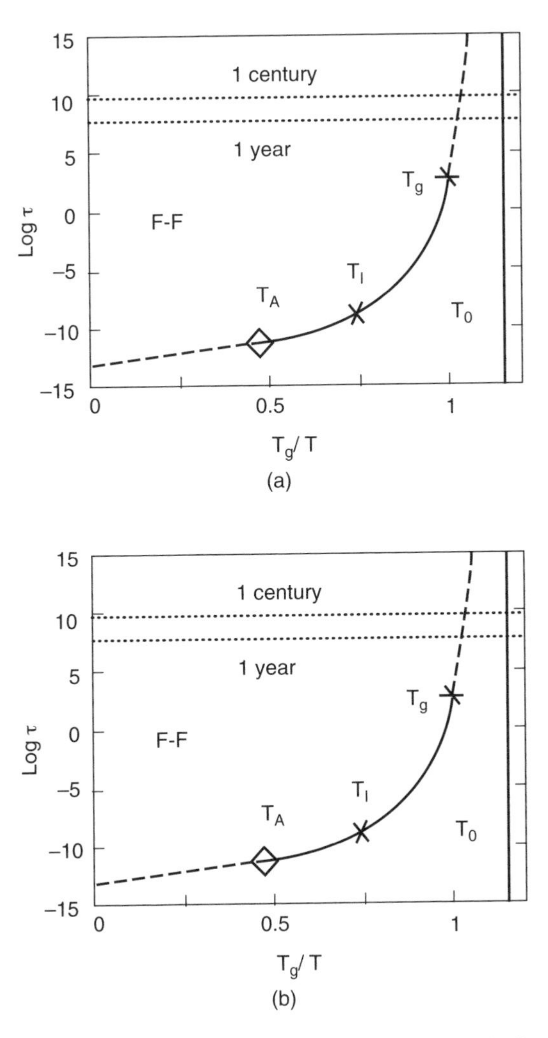

Figure 20. (a) Structural relaxation time τ calculated from the generalized entropy theory for constant pressure ($P = 1$ atm; 0.101325 MPa) high molar mass F-S polymer fluids as a function of the reduced inverse temperature T_g/T. The Arrhenius temperature T_A, crossover temperature T_I, and glass transition temperature T_g are denoted by the same symbols as in Figs. 6 and 7. The remaining characteristic temperature T_0 is indicated as a solid vertical line. Dashed lines refer to high temperature regime $T > T_A$, where an Arrhenius relation $\tau \approx \exp(\beta\Delta\mu)$ applies, and to very low temperatures $T < T_g$ where τ becomes astronomical. The high temperature limiting relaxation time τ_o is taken in these calculations to have the typical order of magnitude $\tau_o = 10^{-13}$ s. (b) Same as (a) but for F-F polymer fluids. (c) Structural relaxation time τ of (a) and (b) (for constant pressure ($P = 1$ atm; 0.101325 MPa) high molar mass F-S and F-F polymer fluids, respectively), replotted as the ratio $\tau/\tau(T_I)$ versus the modified temperature variable $(T_g/T)(T_I - T)/(T_I - T_g)$, where T_g denotes the glass transition temperature and T_I designates the crossover temperature separating the low and high temperature regimes of glass formation. (Part (c) used with permission from J. Dudowicz, K. F. Freed, and J. F. Douglas, *Journal of Chemical Physics* **124**, 064901 (2006). Copyright © 2006 American Institute of Physics.)

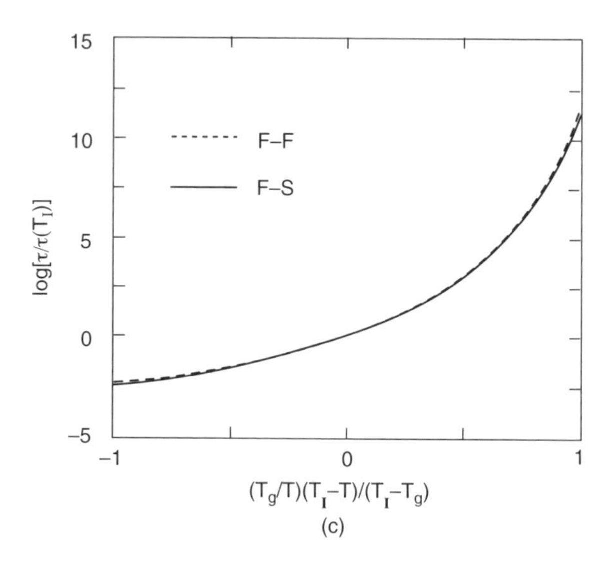

Figure 20. *(Continued)*

equaling 2.08 ± 0.01 and 2.78 ± 0.01 for the F-F and F-S polymer classes, respectively, and with $T_{mc}^{exp} \equiv T_I$. These exponent estimates are consistent with the literature values that lie in the broad range between 2 and 4 for a wide variety of glass-forming liquids (ionic, metallic, organic, water, etc.) [14, 20, 171, 172]. Of course, the theoretical meaning of these apparent exponents is debatable.

We next show that the difference in the temperature dependence of τ between strong and fragile glass-forming polymers can be largely *eliminated* by introducing a modified reduced temperature description involving the variable $T_I/(T_I - T_g)$ that determines the breadth of glass formation. Rössler et al. [142] have found that the temperature dependence of the viscosity $\eta(T_g/T)$ has a nearly universal representation for all fragile and strong polymer fluids when η is normalized by $\eta(T_x)$ and T_g/T is multiplied by $(T_x - T)/(T_x - T_g)$, where the crossover temperature T_x is close to the fitted mode coupling temperature T_{mc}^{exp}. Correspondingly, we apply similar scalings to our computed curves for $\tau(T_g/T)$ in Figs. 20a and 20b by identifying T_{mc}^{exp} with the crossover temperature T_I. Specifically, Fig. 20c presents the normalized relaxation time $\tau/\tau(T_I)$ versus $(T_g/T)(T_I - T)/(T_I - T_g)$. The resulting curves for the F-S and F-F polymer classes are almost indistinguishable from each other over the entire glass-formation temperature range $T_g < T < T_A$. The reduction in Fig. 20c provides theoretical support for the empirical scaling introduced by Rössler et al. [142].

As mentioned in the introduction, the vanishing of s_c and the corresponding divergence of τ in the entropy theory of glass formation appear upon extrapolating a high temperature series expansion theory to low temperatures and therefore have some quantitative uncertainties. If s_c saturates to a constant value at low temperatures, the computed τ does not diverge, but instead reverts back to an Arrhenius temperature dependence. (This Arrhenius behavior emerges directly from Eq. (33) if s_c^*/s_c becomes temperature independent at very low temperatures.) A return to an Arrhenius dependence of τ in glass-forming liquids below T_g has been claimed in a number of investigations [173], and particularly convincing evidence for this phenomenon is provided by O'Connell and McKenna [173] for polycarbonate at low temperatures.

Despite experimental [173] and computational [73] studies supporting a saturation of s_c to a nonzero constant below T_g, we must appreciate the difficulty in equilibrating polymer fluids below T_g, so that a true divergence of τ cannot be excluded. For instance, a divergent dielectric susceptibility has been reported by Menon and Nagel [174] upon approaching the VFTH temperature (corresponding to T_0 in the entropy theory). Given this uncertainty regarding the nature of relaxation at temperatures below T_g because of the extremely large relaxation times, our predictions of the entropy theory of glass formation in Figs. 20a and 20b focus on the temperature range $T_g < T < T_A$, where an equilibrium theory is a reasonable tool to describe the dynamics of glass-forming liquids. We leave the temperature regime below T_g to the same philosophers who calculated the number of angels that could dance on the head of a pin.

X. INFLUENCE OF PRESSURE ON GLASS FORMATION

It is often possible to induce glass formation through a rapid increase of pressure at constant temperature. While this path to glass formation is much less investigated experimentally, this glass-formation process can be analyzed with the combined LCT–AG theories, as we briefly describe.

Interest in the pressure dependence of structural relaxation in fluids has been stimulated by recent applications [175, 176] of a simple pressure analogue of the VFTH equation for the relaxation time τ at a constant pressure P to the analysis of experimental data at variable pressures. Specifically, $\tau(P)$ for both polymer and small molecule fluids has been found to extrapolate to infinity at a critical pressure P_o, and this divergence takes the form of an essential singularity,

$$\tau(P) = \tau(P = 1\text{atm}) \exp[aP/(P_o - P)] \tag{49}$$

where a is a constant at a given temperature. We briefly examine the consistency of this empirical expression for $\tau(P)$ with the entropy theory of glass formation as a severe test of our theory since Eq. (49) is not obviously consistent with our theory.

The relaxation time at a fixed temperature T and variable pressure P can be expressed as a generalization of Eq. (33),

$$\tau = \tau_o \exp(\beta \mathcal{E}_{AG}) \tag{50}$$

with $\mathcal{E}_{AG}$ written equivalently as

$$\mathcal{E}_{AG}(P) = \Delta\mu[s_c^*/s_{c,o}][s_{c,o}/s_c(T,P)] \tag{51}$$

where $s_{c,o} \equiv s_{c,o}(T)$ is the site configurational entropy under the standard condition of atmospheric pressure (1 atm) and where a weak pressure dependence of $\Delta\mu$ and s_c^* is neglected [84, 177]. Equations (50) and (51) can be rearranged as

$$\tau(P) = \tau(P = 1\text{atm}) \exp[\beta \mathcal{E}_{AG}(P = 1\text{atm})\delta s_c] \tag{52}$$

$$\delta s_c \equiv \frac{s_{c,o}T}{s_c(T,P)T} - 1 = \frac{s_{c,o}}{s_c(T,P)} - 1 \tag{53}$$

Consistency of Eq. (52) with the empirical correlation for $\tau(P)$ in Eq. (49) demands that δs_c is inversely proportional to the reduced pressure $\delta P = |P - P_o|/P$.

Figure 21 exhibits δs_c (for an isothermal F-S melt) as a linear function of $1/\delta P$, thereby establishing this consistency to a high degree of approximation. Although the pressure analogue of the VFTH equation has first been discovered experimentally, the origin of this relation naturally follows from the entropy theory. The inset to Fig. 21 presents the Vogel pressure P_o as scaling linearly with temperature, but generally P_o is also a function of ϵ, E_b, E_s, and monomer structure [178]. A similar approach should apply well for glass formation at constant volume, where δP is replaced by the excess free volume δv as the control parameter, but this case is not pursued here [179].

We also examine how pressure mediates the fragility of glass-forming liquids and how the predicted changes in fragility from our entropy theory compare to recent measurements of fragility changes with pressure. McKenna and co-workers [173] find that PS becomes less fragile at elevated pressures, and the same trend emerges for a wide range of nonassociating glass formers from more recent studies [180, 181] by Roland and co-workers. These results are now compared with the corresponding LCT calculations for F-S class polymers. Figure 22 presents the calculated s_c (normalized by its maximum value s_c^*) for the high molar mass ($M = 40001$) F-S polymers as a function of the reduced temperature $\delta T = (T - T_0)/T_0$, demonstrating that a higher pressure leads to a weaker temperature dependence of $s_c(T)/s_c^*$, especially in the high temperature regime $T_I < T < T_A$. This reduced temperature dependence of $s_c(T)$ at elevated pressures should affect fragility, and, indeed, the fragility K_s of high

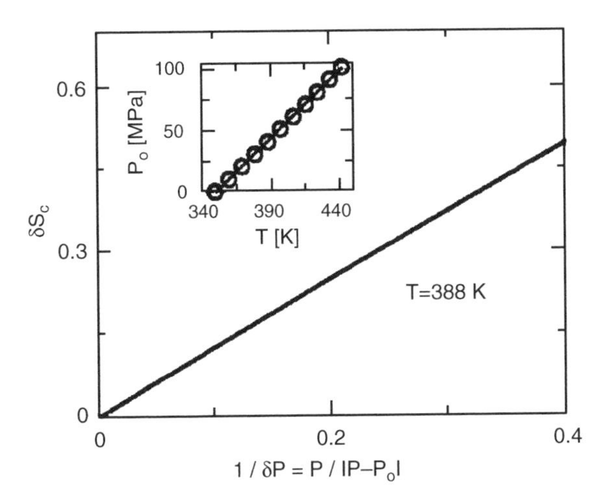

Figure 21. LCT computations for the reduced configurational entropy δs_c defined by Eq. (53) as a function of the reciprocal of the reduced pressure $\delta P = |P - P_o|/P$ (where P_o denotes the Vogel pressure) for high molar mass ($M = 40001$) F-S polymer fluid at fixed temperature $T = 388$ K. The inset illustrates the temperature dependence of P_o (symbols) and the line represents a fit to the data points, using $P_o = a + bT$ with $a = -378.6$ MPa and $b = 1.0825$ MPa/K. (Used with permission from J. Dudowicz, K. F. Freed, and J. F. Douglas, *Journal of Chemical Physics* **123**, 111102 (2005). Copyright © 2005 American Institute of Physics.)

molar mass F-S polymers is computed as 0.27 for $P = 240$ atm compared to 0.36 for $P = 1$ atm (see Fig. 12). A similar trend of decreasing fragility with pressure can be derived from Fig. 7. The fragility parameter K_s of the low temperature regime of glass formation is defined by

$$K_s = (s_c T/\delta T)/(\Delta \mu s_c^*/k_B), \quad T_g < T < T_I \tag{54}$$

in Section V and applies to the temperature range near T_g where the effects of pressure on s_c are relatively small compared to the high temperature regime of glass formation where the influence of pressure becomes appreciable (see Fig. 22). Hence, it is desirable to introduce separate definitions of fragility in the high ($T_I < T < T_A$) and low ($T_g < T < T_I$) temperature regimes of glass formation. Specifically, the parameter C_s defined as the coefficient in the parabolic dependence of $z^* \equiv s_c^*/s_c$ on the reduced temperature T_A (see Eq. (36)),

$$z^* - 1 = C_s \left[\frac{|T - T_A|}{T_A} \right]^2, \quad T_I < T < T_A \tag{55}$$

may serve as a useful measure of fragility in the *high temperature regime*, complementing the definition of K_s. The changes in fragility at high temperatures

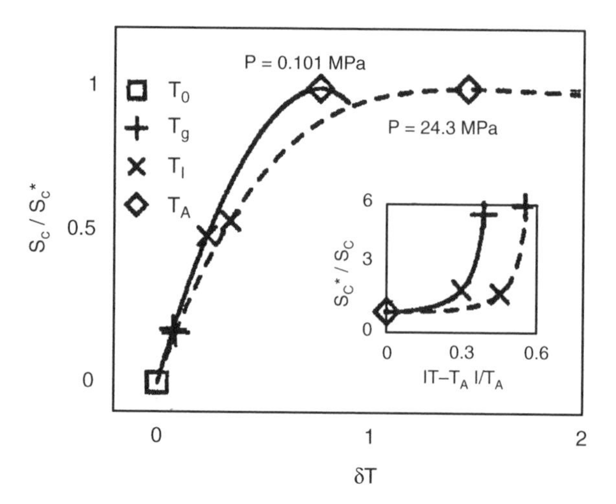

Figure 22. The configurational entropy s_c per lattice site as calculated from the LCT for a constant pressure, high molar mass ($M = 40001$) F-S polymer melt as a function of the reduced temperature $\delta T \equiv (T - T_0)/T_0$, defined relative to the ideal glass transition temperature T_0 at which s_c extrapolates to zero. The specific entropy is normalized by its maximum value $s_c^* \equiv s_c(T = T_A)$, as in Fig. 6. Solid and dashed curves refer to pressures of $P = 1$ atm (0.101325 MPa) and $P = 240$ atm (24.3 MPa), respectively. The characteristic temperatures of glass formation, the ideal glass transition temperature T_0, the glass transition temperature T_g, the crossover temperature T_I, and the Arrhenius temperature T_A are indicated in the figure. The inset presents the LCT estimates for the size $z^* = s_c^*/s_c$ of the CRR in the same system as a function of the reduced temperature $\delta T_A \equiv |T - T_A|/T_A$. Solid and dashed curves in the inset correspond to pressures of $P = 1$ atm (0.101325 MPa) and $P = 240$ atm (24.3 MPa), respectively. (Used with permission from J. Dudowicz, K. F. Freed, and J. F. Douglas, *Journal of Physical Chemistry B* **109**, 21350 (2005). Copyright © 2005, American Chemical Society.)

are quantified in the inset to Fig. 22. Table I indicates that $C_s = 7.10$ for high molar mass F-S polymers at $P = 1$ atm, while a significantly smaller $C_s = 0.69$ is obtained for this system for $P = 240$ atm. (The constant of proportionality in a relation like Eq. (55) has been advocated as a measure of fragility in spin models exhibiting glass formation [182], where the power in the reduced temperature $|T - T_A|/T_A$ is found to be somewhat larger than 2.) Note that Eq. (55) is compatible with recent experimental correlations for the reduced activation energy z^* of diverse fluids by Kivelson et al. [13].

XI. ALTERNATIVE ACTIVATION ENERGY THEORIES OF DYNAMICS IN GLASS-FORMING LIQUIDS

An obvious shortcoming of the entropy theory of glass formation is the fact that s_c is not directly measurable. Section VI describes how the relative specific volume δv can serve as surrogate for s_c under restricted circumstances. Given the results of the previous section, it is natural to inquire into whether other thermodynamic properties exhibit useful approximate relations to s_c.

Schweizer and Saltzman [39, 40] (SS) have recently developed an ambitious treatment of glass formation based on a thermodynamic barrier concept in which the barrier height $\mathcal{E}$ is expressed in terms of the fluid isothermal compressibility κ_T rather than the configurational entropy. (This creative approach incorporates ideas taken from a combination of mode coupling theory, density functional theory, and activated rate theory in order to provide an analytical estimate [39, 40] for $\mathcal{E}$.) Other reasonably successful phenomenological descriptions of relaxation in glass-forming liquids propose [57] a proportionality between $\mathcal{E}$ and the high frequency shear modulus G_∞ (the "shoving model" [57]). The present section describes these alternative "activation energy" theories of glass dynamics and their relation to our entropy theory of glass formation. In particular, our estimate of the entropic barrier height $\mathcal{E}_{AG}$ is compared below with the corresponding estimate [39, 40] of $\mathcal{E}$ by SS, which leads to an approximate scaling relation between s_c and κ_T. A similar scaling is then derived between s_c and the Debye–Waller factor $\langle u^2 \rangle$, based on an analysis of G_∞.

A. Relation to Schweizer–Saltzman Theory

According to SS, the structural relaxation time of a glass-forming liquid is described by a generalized Arrhenius expression [39, 40],

$$\tau = \tau_A \exp(F_B) \tag{56}$$

where τ_A is the relaxation time at high temperatures, which is given in our notation by

$$\tau_A = \tau_o \exp(\beta\, \Delta\mu) \tag{57}$$

and F_B in Eq. (56) is a dimensionless barrier height that quantifies the increase of τ from its counterpart τ_A for noncollective motion. The barrier height F_B predicted by the SS theory for a *colloidal* glass-forming liquid can be approximated numerically by the inverse compressibility ratio,

$$F_B \approx 1/S(0) = 1/(\kappa_T/\kappa_T^o) = \kappa_T^o/\kappa_T \tag{58}$$

where $\kappa_T^o = 1/\rho k_B T$ denotes the fluid isothermal compressibility of an ideal gas and κ_T refers to the dense fluid. Equation (58) is particularly interesting since it is expressed in terms of the thermodynamic properties of the fluid, and it exhibits no explicit dependence on the parameters characterizing the particles for which the SS theory is developed.

The inverse relation between the fluid activation energy and the isothermal compressibility can naturally be derived by extending prior arguments by Bueche [183] suggesting that the frequency of molecular displacements in dense fluids is governed by *fluctuations* in the local volume near a test particle. Specifically, Bueche predicts that these fluctuations can be described by a Gaussian distribution with the variance governed by the fluid compressibility κ_T (which is proportional to the variance in the density fluctuations) [183]. Combining these arguments with the fluctuation model of structural relaxation by Gordon [184] leads directly to an inverse relation between $\mathcal{E}$ and κ_T, as in Eq. (58). Thus, the relation (58) is highly plausible.

A generalization of the SS method to polymer liquids indicates that the inverse relation between F_B and $S(0)$ only holds *qualitatively* for the coarse grained idealized "thread model" of polymer chains that renders the computations tractable. In this generalized model, the expression for F_B becomes rather complicated, and the result depends on the degree of polymerization and other molecular parameters. Given the above formal arguments relating τ to volume fluctuations, the convincing nature of the derivation of the expression for F_B in the case of colloidal fluids, and experimental evidence indicating that the phenomenology of glass formation is common to both polymeric and colloidal glass formers, we consider Eq. (58) further as an *approximation* [185] for polymeric fluids as well.

Based on comparisons with experimental data at high temperatures, SS introduce the approximate relation for polymer fluids, [39, 40],

$$\kappa_T^o/\kappa_T \approx C_\kappa[(T - T^*)/T]^2\,, \quad T < T^* \tag{59}$$

where T^* and C_κ are constants specified by SS [186]. On the other hand, Eq. (33) implies that the entropy theory counterpart of the barrier height F_B equals $(z^* - 1)(\beta\,\Delta\mu)$ and thus suggests the formal correspondence

$$(z^* - 1)(\beta\,\Delta\mu) \leftrightarrow \kappa_T^o/\kappa_T$$

which indicates a possible interrelation between s_c and κ_T. This correspondence is not pursued here, beyond checking the temperature dependence of $(z^* - 1)(\beta\,\Delta\mu)$ that emerges from generalized entropy theory in the vicinity of T_A. In particular, the form

$$(z^* - 1)(\beta\,\Delta\mu) = (C_s\beta\,\Delta\mu)[(T - T_A)/T_A]^2 \tag{60}$$

which applies over the temperature range $T_A - 100\,\mathrm{K} < T < T_A$, follows directly from Eq. (36). The above analysis suggests that the barrier heights estimated from the LCT and the SS theory scale with temperature near T_A in a qualitatively consistent fashion. Schweizer and Saltzman also find [39, 40] that their expression for F_B for polymer liquids accords with the Rössler scaling [142], where the approximation introduced by Eq. (58) is *not invoked.* An expression mathematically equivalent to Eq. (60), except for a larger exponent (8/3 rather than 2), has been proposed based on the "frustration-limited cluster" model of Kivelson et al. [13]. This model, like the AG model, attributes the increasing barrier height in cooled liquids to the growing size of dynamic clusters in cooled liquids. Overall, this brief comparison seems to indicate that the SS theory has much in common with the generalized entropy theory of glass formation, and this interrelation deserves a more detailed investigation. We find the ability of the SS theory to relate the kinetic barrier F_B to long wavelength collective variables like $S(0)$ to be an extremely encouraging aspect of this approach for treating the molecular dynamics of glass-forming and other complex fluids.

B. Relation of Entropy Theory to Elastic Modulus Model of Structural Relaxation

Another class of thermodynamic barrier theories focuses on the large increases in the elastic constants that accompany glass formation. (These theoretical approaches seem especially appropriate to polymer fluids below the crossover temperature T_I.) In particular, the barrier height $\mathcal{E}$ governing particle displacement in the shoving model [57] is taken to be on the order of the elastic energy $G_\infty V_o$ required to displace a particle on a scale comparable to the interparticle distance,

$$\mathcal{E} \approx G_\infty V_o \tag{61}$$

where V_o is an unspecified "critical" volume of molecular dimensions. Note that if we further invoke the approximation [187] $G_\infty \approx 1/\kappa_T$, then Eq. (61) exhibits the same scaling with κ_T, as suggested by SS (i.e., $\mathcal{E} \approx 1/\kappa_T$).

A further development is possible by noting that the high frequency shear modulus G_∞ is related to the mean square particle displacement $\langle u^2 \rangle$ of caged fluid particles (monomers) that are transiently localized on time scales ranging between an average molecular collision time and the structural relaxation time τ. Specifically, if the viscoelasticity of a supercooled liquid is approximated below T_I by a simple Maxwell model in conjunction with a Langevin model for Brownian motion, then $\langle u^2 \rangle$ is given by [188]

$$\langle u^2 \rangle = 2\beta / \pi R G_\infty \tag{62}$$

where R denotes the particle (monomer) radius. Although this is a rather idealized model, we anticipate that the inverse scaling between $\langle u^2 \rangle$ and G_∞

should be preserved in more complex treatments of the viscoelasticity of glass-forming fluids. The formal correspondence between $\mathcal{E}_{AG} \approx (s_c^*/s_c)\Delta\mu$ and $\mathcal{E}$ of Eq. (61) then implies the extraordinary scaling relation

$$s_c T \sim \langle u^2 \rangle \tag{63}$$

between the configurational entropy s_c and the Debye–Waller factor $\langle u^2 \rangle$ characterizing the amplitude of molecular displacements within the fluid at very short times (e.g., typically nanoseconds in elastic neutron scattering measurements). Both the landscape configurational energy $s_{c,L}T$ and $\langle u^2 \rangle$ have been found [152] to scale linearly with $\delta T = (T - T_0)/T_0$, and the landscape entropy equivalent form of Eq. (63), $s_{c,L}T \sim \langle u^2 \rangle$, has been suggested previously based on molecular dynamics simulations [152]. A direct test of the relation (63) is unavailable at the moment. Starr et al. [152] note that $\langle u^2 \rangle$ can be considered as a molecular scale measure of excess free volume, so that the common extrapolation of $s_c T$ and $\langle u^2 \rangle$ to zero at T_0 is quite natural. The combination of the AG equation (33) with Eq. (63) yields the Buchenau relation [58] between $\langle u^2 \rangle$ and τ:

$$\tau \approx \tau_o \exp(\langle u^2 \rangle_o / \langle u^2 \rangle) \tag{64}$$

where $\langle u^2 \rangle_o$ is a constant. The relation in Eq. (64) is found to describe relaxation data for a variety of glass-forming liquids [58, 189, 190] and has been verified in molecular dynamics simulations of a supercooled polymer melt [152].

In view of Eq. (45) and the approximate Maxwell scaling relation $\eta \sim G_\infty \tau$, it is also possible to introduce another measure m_{u^2} of fragility [190]:

$$m_{u^2} \equiv \left. \frac{\partial(\langle u^2 \rangle_o / \langle u^2 \rangle)}{\partial(T_g/T)} \right|_{T=T_g} \tag{65}$$

The application of this estimate of fragility to data for $\langle u^2 \rangle$ from incoherent neutron scattering has been reported to correlate well with estimates of fragility obtained from viscosity measurements [190]. This consistency is rather remarkable given the vastly different time scales involved in neutron scattering and viscosity measurements.

C. Generalized Lindemann Criteria for the Stages of Glass Formation

Our identification of the Buchenau relation in Eq. (64) with the AG equation for τ further implies that the characteristic temperatures of glass formation (T_0, T_g, T_I, T_A) have their counterparts in characteristic values of $\langle u^2 \rangle$ and that the Lindemann criterion (see Section VI) can be generalized to describe successive stages of glass formation. Dividing the arguments of the exponentials defining

τ in Eqs. (33) and (64) by their values at T_g leads to the following consistency conditions:

$$\frac{\langle u^2 \rangle_{T_I}}{\langle u^2 \rangle_{T_g}} = \left(\frac{T_I}{T_g}\right) \frac{z^*(T = T_g)}{z^*(T = T_I)} \tag{66}$$

and

$$\frac{\langle u^2 \rangle_{T_A}}{\langle u^2 \rangle_{T_g}} = \left(\frac{T_A}{T_g}\right) z^*(T = T_g) \tag{67}$$

where $z^*(T = T_A) \equiv 1$. For high molar mass F-S polymers, where $T_A/T_g = 1.64$ and $z^*(T = T_g) = 5.53$, the root-mean-square "cage" size $\langle u^2 \rangle^{1/2}$ at the Arrhenius temperature T_A is found to be about 3.0 times its value at T_g:

$$(\langle u^2 \rangle_{T_A})^{1/2} \approx 3.0(\langle u^2 \rangle_{T_g})^{1/2} \tag{68}$$

This condition accords well with the numerical estimate of La Violette and Stillinger [130], who report that the liquid first becomes unstable to freezing when $\langle u^2 \rangle^{1/2}$ is about 3 times the Lindemann value of $\langle u^2 \rangle^{1/2}$ characterizing the instability of the solid toward melting. Thus, both the onset and the end of the glass transformation process can symmetrically be described by *generalized Lindemann criteria*. The middle of the glass-transition range is specified by an intermediate value of $\langle u^2 \rangle$. Substituting the values of T_I/T_g and $z^* = 2.02$ for F-S polymers (see Table I) into Eq. (66) yields

$$(\langle u^2 \rangle_{T_I})^{1/2} = 1.77(\langle u^2 \rangle_{T_g})^{1/2} \tag{69}$$

The estimate of $(\langle u^2 \rangle_{T_I})^{1/2}$ lies nearly in the center of the range between $(\langle u^2 \rangle_{T_g})^{1/2}$ and $(\langle u^2 \rangle_{T_A})^{1/2}$. The limit $T \to T_0$ corresponds to the vanishing of s_c and thus to the vanishing of $\langle u^2 \rangle$. This extrapolated vanishing of $\langle u^2 \rangle$ as temperature approaches the VFTH temperature has recently been observed in MD simulations of polymer melts [152].

XII. COOPERATIVELY REARRANGING REGIONS, EQUILIBRIUM POLYMERIZATION, AND DYNAMIC HETEROGENEITY

While the combination of the LCT for polymer glasses and the postulates of the Adam–Gibbs model for relaxation in cooled liquids captures many aspects of glass formation, the resulting entropy theory leaves unresolved certain microscopic features of glass formation. For instance, many experimental and computational studies indicate the presence in the cooled fluid of transient domains with high and low particle mobilities relative to fluids exhibiting simple Brownian motion [17, 35, 113, 114, 117, 178, 191–204]. An elucidation of the origin of these dynamically heterogeneous structures and their impact on fluid relaxation remains a challenge for any theory of glass formation. Much of the discussion in this section represents a speculative attempt at accounting for this dynamic heterogeneity and its ramifications for fluid relaxation in a framework that is consistent with the entropy theory.

Mobile particle clusters have been observed in simulations of cooled glass-forming liquids [113, 114, 199, 200] and in concentrated colloidal fluids [195, 196] that form glasses. In particular, simulations of glass-forming binary LJ fluids, water at low temperatures, and model polymer glass-forming liquids reveal the existence of *polymer-like* clusters of mobile particles that form and disintegrate in a state of dynamic equilibrium [113–116]. Mobile polymeric particle clusters have also been identified in suspensions of spherical colloid particles at high concentrations [195–197]. More quantitatively, Donati et al. [113] find that these transient polymeric structures have a nearly exponential size distribution, a characteristic feature of equilibrium polymerization systems [205]. Moreover, the apparent activation energy $\mathcal{E}$ for relaxation in these fluids has been claimed to be proportional to the mass of the mobile clusters and to scale [115, 116] inversely to the landscape configurational entropy $s_{c,L}$. A similar inverse scaling between the average degree of polymerization L and the configurational site entropy $s_c(T)$ applies for "living" equilibrium polymers or, equivalently, for equilibrium polymerization with a chemical initiator [205]. These findings are all consistent with considering the cooperatively rearranging regions of AG as being analogous to equilibrium polymers [113, 115, 116, 205].

The analogy between the CRR and equilibrium polymers provides a potential basis for understanding dynamic heterogeneity in cooled liquids, the factors controlling the fragility of these fluids, and the origin of nonexponential relaxation as arising from the inherent polydispersity of the mobile and immobile particle clusters formed at equilibrium [206]. In addition, the origin of physical aging [207, 208] and of nonlinear rheological properties [209] (e.g., shear thinning, normal force effects) of glass-forming liquids can be rationalized by the slow approach of the polymeric structures to their equilibrium population and by the disruption of these transient clusters under flow, respectively. While the identification of the CRR of the entropy theory with equilibrium polymers is highly suggestive, it is necessary to test whether the treatment of the CRR polymerization process by equilibrium polymerization theory is consistent with the thermodynamic description of glass formation provided by the entropy theory.

Polymeric excitation structures [210], characterized by equilibrium polymerization transitions, are known to drive thermodynamic transitions in many model systems (XY models [210, 211] of superfluidity [212], superconductivity [213], dislocation models of melting [214], sulfur polymerization [215, 216], liquid crystal ordering [217, 218], etc.). These transitions may appear as first or second order phase transitions [210–220] or as rounded [205, 221] (i.e., infinite order) phase transitions. The broad nature of glass formation emerging from the LCT and experiments clearly points to such a rounded transition, as emphasized by Gordon et al. [71]. The essential characteristics of this kind of rounded transition are apparent from the temperature variation of the configurational entropy in the simplest models of equilibrium polymerization. In particular, we briefly recall some of our previous results for a model in which each particle can freely associate in a solvent to form linear chains, a model which is termed the free association ($\mathcal{FA}$) model. This comparison allows us to examine the second Adam–Gibbs hypothesis (see Section II) of an inverse relation between the configurational entropy and the mass of the dynamic clusters in the cooled liquid.

A. Test of the Adam–Gibbs Second Hypothesis for Equilibrium Polymerization

Figure 23 presents the computed temperature dependence of the dimensionless configurational entropy $s_c^{(\mathrm{red})}$ (for the ($\mathcal{FA}$) model) defined as

$$s_c^{(\mathrm{red})} = [s_c(T) - s_c^{(o)}]/[s_c^{(\infty)} - s_c^{(o)}] \tag{70}$$

where $s_c(T)$ is the system's configurational entropy (per lattice site) at a given temperature T, and $s_c^{(o)}$ and $s_c^{(\infty)}$ denote the low and high temperature limits of $s_c(T)$, respectively, where s_c saturates to constants. The normalization in Eq. (70) ensures that $s_c^{(\mathrm{red})}$ varies from zero to unity upon heating. The decreasing reduced configurational entropy $s_c^{(\mathrm{red})}$ in Fig. 23 upon cooling is analogous to the configurational entropy $s_c(T)$ of glass-forming liquids in Fig. 6, and the average degree of polymerization L is nearly inversely proportional to $s_c^{(\mathrm{red})}$ (see the inset to Fig. 23), in close agreement with the second Adam–Gibbs hypothesis [48]. Moreover, the variation of L with $s_c^{(\mathrm{red})}$ is similar to the variation of z^* with $s_c(T)$ for glass-forming liquids, consistent with our identification of the CRR with equilibrium polymers. In addition, a striking inverse relation $L \sim 1/s_c^{(\mathrm{red})}$ has been found previously in the case of living polymerization, independent of initiator concentration [205]. Thus, the inverse proportionality between the reduced configurational entropy and the average degree of polymerization of the equilibrium polymers seems to be a general property of particles associating into clusters at equilibrium, and the second AG hypothesis seems to be quite plausible.

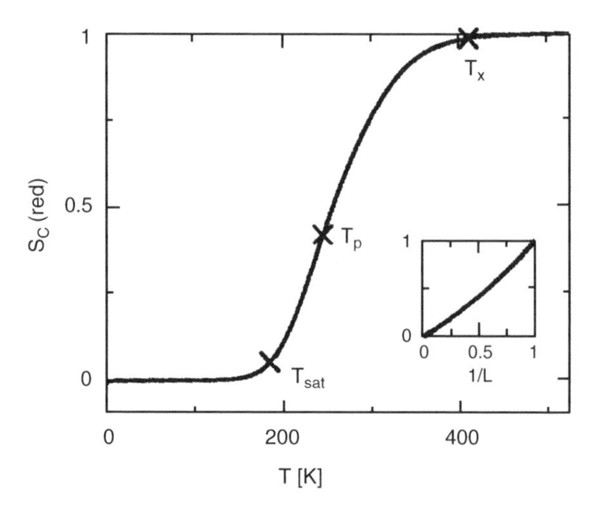

Figure 23. Temperature dependence of the reduced configurational entropy per unit site $s_c^{(\mathrm{red})}$ (defined by Eq. (70)) for the free association equilibrium polymerization model in the infinite pressure limit. The values of the enthalpy $\Delta h_p = -35$ KJ/mol and entropy $\Delta s_p = -105$ J/(mol K) of polymerization are identical to those used in our extensive studies of equilibrium polymerization,[205, 222, 223], and the initial monomer concentration ϕ_1^o is taken as $\phi_1^o = 0.1$. The crossover temperature T_x is identified [205] as the temperature at which the fraction Φ of monomers converted into polymers is 5%, while the saturation temperature T_{sat} refers[205] to the temperature where the entropy of the living polymer solution approaches within 5% of its limiting low temperature value $s_c^{(o)}$. The inflection temperature T_Φ and the polymerization transition temperature T_p are defined as the temperatures where $(\partial^2\Phi/\partial T^2)|_{\phi_1^o} = 0$ and where the specific heat $C_V(T)$ exhibits the maximum, respectively [205]. The inset shows that the reduced configurational entropy $s_c^{(\mathrm{red})}$ is roughly inversely proportional to the average degree of polymerization L.

B. Characteristic Temperatures of Polymerization and Glass Formation, Transition Broadening, and Fragility

The thermodynamic transition predicted by equilibrium polymerization theory [205, 222] in Fig. 23 is broad and can be described by three main characteristic temperatures: an onset temperature T_x where polymers first begin to form, a polymerization temperature where the (configurational) specific heat has a maximum, and a saturation temperature T_{sat} where s_c nearly saturates to its low temperature limit [205, 223]. The polymerization temperature T_p represents the demarcation between a monomer dominated regime ($L \approx 1$–2) at high temperatures and a polymeric, low temperature regime ($L \gg 2$). No phase transition of finite order (first or second order transition) occurs in the $\mathcal{FA}$ model since all derivatives of the free energy are finite at finite temperatures. In other words, the behavior in Fig. 23 is a classic example of a "rounded" thermodynamic transition [223]. The multiplicity of characteristic temperatures in glass formation is certainly suggestive of a similar rounded thermodynamic transition.

The wide temperature range between T_{sat} and T_x is a measure of the extent of transition rounding in the $\mathcal{FA}$ model [223], and this transition rounding has its counterpart as the fragility for glass-forming liquids, where the ratios of characteristic temperatures are taken as measures of the transition "sharpness." Specifically, the onset temperature T_x, the polymerization temperature T_p, and the saturation temperature T_{sat} qualitatively correspond, respectively, to the characteristic temperatures T_A, T_I, and T_0 of the entropy theory of glass formation. More general models of equilibrium polymerization involve constraints on how the clusters grow (initiation of growth from the chain ends [224], thermal activation [223], etc.), which, in turn, affect transition sharpness (i.e., rate of change $s_c^{(red)}(T)$ with temperature or the relative gaps between the characteristic temperatures T_x, T_p, and T_{sat}). Thus, we can adjust the "fragility" of the equilibrium polymerization system by varying the constraints acting on the polymerization process. Specifically, the rate of change of the viscosity of equilibrium polymer solutions (e.g., poly(α-methylstyrene)) with temperature can be continuously tuned, for example, by varying the initiator concentration [224, 225], a parameter controlling the sharpness of this particular type of equilibrium polymerization transition [221, 223]. The suggested correspondence between equilibrium polymerization and glass formation provides clear hints of the physical factors that affect the variation of fragility in glass-forming liquids. For example, energetically unfavorable (unstable) regions that form spontaneously by thermal fluctuations could play the role of an initiator in catalyzing collective liquid motion, as suggested previously for water [226]. We anticipate this viewpoint will be useful in comprehending the dynamical meaning of fragility variations in glass-forming liquids at a more molecular scale.

C. Size of Cooperatively Rearranging Regions and the Average Degree of Polymerization: Evidence for Universality at the Crossover Temperature T_I

The proposed analogy between the CRRs and equilibrium polymers has implications for the number of particles involved in these structures and their spatial dimensions. The size z^* of the cooperatively rearranging regions near the crossover temperature separating the high and low temperature regimes of glass formation (our T_I) is about 2 for fragile glass-forming liquids [15]. Correspondingly, the average polymerization index L at T_p happens to be close to 2 in the $\mathcal{FA}$ model discussed earlier. This value compares well with the average polymer "string length" $L \approx 2.2$ of mobile particles (calculated from MD simulation data for binary LJ glass-forming liquids) [113, 114] in the vicinity of T_{mc}, which, as discussed in Section V, is a specific estimate of the crossover temperature T_I of glass formation. Recent experiments [227] and simulations [228] of associating dipolar fluids also reveal the presence of a polymerization transition involving the formation of equilibrium polymer chains where the magnitude of L at the polymerization temperature T_p is found in the comparable range between 2 and 4 (in both three and two spatial dimensions). The near universality of the extent of cooperative motion (z^*) near the crossover temperature of glass formation [15], the direct observation of transient polymer clustering in simulations of glass-forming liquids [113], and the corresponding predictions from theories and simulations of fluids exhibiting equilibrium polymerization [223, 227, 228] provide good further supporting evidence that the cooperative motion in structural glasses involves a kind of equilibrium polymerization.

The suggested correspondence between equilibrium polymers and the CRR of the Adam–Gibbs model allows us to estimate the spatial dimensions of these hypothetical clusters. Taking the size of a molecule or monomer to be on the order of 0.5–1 nm in a typical glass-forming liquid, such as orthoterphenyl [127], the average size of a CRR at the glass transition should likewise have nanometric dimensions. Specifically, if we assume that the root-mean-square end-to-end distance R_{rms} of the CRR grows with the square root of the degree of polymerization (i.e., $R_{rms} \sim (z^*)^{1/2}$), then this measure of average CRR size is roughly in the range from 1 to 2 nm. This estimate presumes that z^* at T_g is about 4–5, depending on the polymer class (see Table I). Of course, the random walk scaling predictions [113] represent an idealization for such short chains, and a worm-like chain model might be more plausible for the mean dimensions of the CRR polymeric clusters. An upper bound for R_{rms} can be inferred by assuming rod-like CRR chains, ranging from 2 to 5 nm. A realistic value of R_{rms} should then lie midway between the upper and lower bounds, so that 2–3 nm is probably a typical order of magnitude for $0R_{rms}$ to be expected, based on the LCT calculations and the analogy of CRR with equilibrium polymers. This estimate is quite consistent with experimental estimates for the spatial dimensions of the CRR [127, 191, 229]. However, these estimates for the size of the CRR refer to *average cluster dimensions*, and much larger clusters may occur due to the intrinsic mass polydispersity of equilibrium polymers. Moreover, since different measurements are governed by *different averages* of the cluster polydispersity, experimental determinations of the CRR can be anticipated to vary somewhat with the measurement method, a well known phenomenon for the properties of polydisperse polymer solutions.

Garrahan and Chandler [230] have recently attempted to rationalize the string-like motion in supercooled liquids based on a completely different concept of "dynamic facilitation," derived from the study of magnetic spin models originally developed by Fredrickson and Anderson [231]. Although these spin models seem to exhibit dynamic heterogeneity of some kind and slow relaxation processes, the slowing down of the dynamics in these models is entirely *decoupled* from the spin model's thermodynamics [116, 230]. In view of this decoupling and the obscurity of any relation between these spin models and real glass-forming fluids, we conclude that this class of models is entirely at odds with the AG theory and the extensive phenomenology supporting the AG picture of structural glass formation.

D. Implications of Equilibrium Polymerization for Understanding Relaxation Processes in Glass-Forming Liquids

The growth of transient polymeric clusters in supercooled liquids should produce nontrivial changes in their viscoelastic relaxation, just as in the formation of thermally reversible gels of associating molecules. Indeed, the appearance of anomalous viscoelastic properties has led to some of the earliest indications of dynamic heterogeneity in glass-forming liquids [206]. Importantly, the occurrence of clustering at *equilibrium* in liquids suggests a natural mechanism for understanding the universal occurrence of stretched exponential relaxation ($\Psi(t) \sim \exp(-t/\tau)^{\beta}$) that is generally observed in glass-forming liquids [9, 206]. Douglas and Hubbard [206] argue that this stretched exponential relaxation in glass-forming liquids derives from polydispersity of the dynamic clusters, where individual clusters relax exponentially with a rate governed by their size and where β relates to the cluster geometry.

The bifurcation into "fast" and "slow" structural relaxation processes can also be qualitatively understood from this analogy with equilibrium polymerization. Since a "minimal polymer" is arguably composed of two segments (i.e., $z^* = 2$), the emergence of truly polymeric clusters should first appear in the relaxation data for $T \leq T_{\mathrm{I}}$ since the "polymerization index" z^* for cooperative motion exceeds 2 for temperatures near T_{I} (see Table I). A bifurcation [3, 7, 9, 232, 233] of the relaxation function into separate "branches" should then occur near T_{I}, with one branch corresponding to relaxation processes that are dominated by the dynamic polymeric structures and the other branch arising from relaxation processes governed by the monomers that coexist in dynamic equilibrium. Such a bifurcation has often been observed in glass-forming liquids at a temperature near the crossover temperature T_{I} of glass formation and, on the other hand, in associating telechelic fluids [234] and monomeric fluids subjected to irreversible polymerization [235–237]. We anticipate from this physical picture that the stretching exponent β should begin to decrease below T_{A} as the averaging over an increasingly broad cluster size distribution produces progressively larger deviations from the single exponential relaxation characteristic of a homogeneous (unpolymerized) fluid [14, 206]. A breakdown of the Stokes–Einstein relation between diffusion and viscosity is a natural outcome of this emergent dynamic heterogeneity [192, 193, 232, 238, 239]. Many observed features of supercooled fluids are thus potentially comprehensible and consistent with this equilibrium polymerization picture of dynamic heterogeneity in glass-forming fluids. Further support for the analogy between the CCR and equilibrium polymerization is given in Ref. 246.

XIII. CONCLUSIONS

By extending the lattice cluster theory (LCT) to describe glass formation in polymer melts and by adopting venerable concepts from the Gibbs–DiMarzio (GD) and Adam–Gibbs (AG) theories of glass formation, we self-consistently explain the relation between the characteristic temperatures of glass formation within a unified theoretical framework that describes the influence of monomer molecular structure on all properties of glass formation. These characteristic temperatures include the onset temperature T_A for the supercooled regime (below which the structural relaxation time τ no longer displays an Arrhenius temperature dependence), the crossover temperature T_I (separating well defined high and low temperature regimes of glass formation in which both s_c and τ exhibit a rather distinct temperature dependence), the glass transition temperature T_g (associated with fluid jamming), and the ideal glass transition temperature T_0 (at which the fluid configurational entropy s_c extrapolates to zero). While the concepts of the configurational entropy s_c and the "ideal" transition temperature T_0 intrinsically follow from GD theory, the AG postulate of a specific relation between τ and s_c allows the self-consistent definition of both T_A and T_I in terms of the s_c computed from the LCT as a function of monomer molecular structure. This progress in the predictive capacities of the entropy theory of glass formation arises, in part, because the LCT generates analytical expressions for thermodynamic properties, including the effects of short range correlations stemming from chain connectivity, different flexibilities of chain backbone and side groups, and monomer structure, molecular factors that cannot be described by the classic GD theory. Notably, the former GD theory exclusively focuses on computing the temperature T_0, which occurs well below the kinetic T_g, so that the scope of GD theory is much more limited than our entropy theory of glass formation.

Our analytical expression for the temperature dependence of the configurational entropy s_c per site (which strictly is a configurational entropy density) offers the possibility of analyzing the global nature of glass formation over a wide temperature range extending from the onset temperature T_A, where the orientational correlations first start to develop and the complex molecular dynamics of glass formation first appears, to the extrapolated temperature T_0, where the liquid regime ends and solidification is completed either as an equilibrium or nonequilibrium process. While the temperature variation of s_c is roughly linear at low temperatures, this dependence becomes quadratic at high temperatures (see Fig. 6), where s_c achieves a maximum at T_A. (As pressure increases and excess free volume decreases, the maximum in s_c shifts to higher temperatures and finally disappears altogether.) Thus, two different temperature regimes with differing temperature variations appear for s_c below T_A and are separated from each other by a crossover or inflection temperature T_I, defined by an inflection point in the product of $s_c T$ as a function of T. Phenomenologically, T_I is identified with the experimentally determined mode coupling temperature T_{mc}^{exp}, an identification that is supported by the coincidence of computed ratios T_I/T_g from the LCT with literature estimates of T_{mc}^{exp}/T_g and the mode coupling type singular extrapolation of the computed τ at T_I. The remaining characteristic temperature of glass formation, the glass transition temperature T_g ($T_0 < T_g < T_I$), is calculated by applying a Lindemann criterion to the softening transformation in glass-forming liquids (see Section VI). This dynamical definition of T_g asserts that the glass transition T_g corresponds to an isothermal free volume condition having no overt thermodynamic signature. Our estimates of T_g conform well to the phenomenology of glass formation in many systems.

Because our goal lies in investigating the general characteristics of polymer glasses, we introduce a schematic model of glass formation that contains a minimal set of physical parameters (nearest neighbor van der Waals interaction energy ϵ and differing *gauche* energy penalties in the chain backbone (E_b) and the side groups (E_s) that reflect their different rigidities) and that applies simultaneously to three main categories of glass-forming polymers characterized by the relative values of E_b and E_s: chains with a flexible backbone and flexible side groups, chains that have a flexible backbone and rigid side branches, and chains with a relatively stiff backbone and flexible side groups. The distinction between these three categories, which are termed the F-F, F-S, and S-F polymer classes, respectively, is inspired by the classification of the relation between monomer structure and polymer fragility by Roland and coworkers [77, 79] and by Colucci and McKenna [78]. Examples of each polymer class are given in Section III. Our calculations for various measures of fragility confirm previous experimental observations, indicating that the presence of bulky, rigid side groups in polymer chains leads to an increased fragility of the glass.

The overall breadth of the temperature range over which glass formation occurs and the rapidity with which the configurational entropy s_c (and fluid fragility) varies with temperature are intimately related. Thus, ratios of the characteristic temperatures of glass formation provide model-independent information about fragility (i.e., larger temperature ratios imply a broader glass transition and stronger glass formation). These characteristic temperatures (T_0, T_g, T_I, T_A) are evaluated for the model F-F and F-S classes of polymers to understand general trends in fragility and the dependence of these temperatures on molar mass and pressure. All the ratios T_A/T_0, T_A/T_g, T_A/T_I, and T_I/T_g are found to be larger for the F-F class of polymers than for the F-S class, while the M dependence of these ratios turns out to be weak for both classes. The relatively large temperature ratios indicate that the F-F polymers are stronger glass formers than F-S polymers. A direct calculation of the fragility parameter

K_s (using the entropy theory) for the F-F and F-S classes of polymers leads to remarkable agreement with experimental data and confirms the more fragile nature of F-S polymers. At a structural level, these results imply, in accord with experiment [79], that polymer chains with bulky, stiff side groups have higher fragility than polymers with side groups whose molecular structure more resembles the chain backbone segments and whose rigidity is likewise similar to the rigidity of the backbone.

Further insight into the structural origin of variations of fragility in glass-forming polymer liquids emerges from considering the specific volume v and isothermal compressibility κ_T. These properties exhibit a weaker temperature dependence for the F-F polymer class than for the F-S polymer class. Increasing pressure reduces ϕ_v and diminishes fragility, a behavior that is more dramatic for the F-S polymers at high temperatures, where a separate definition of fragility is introduced for more sensibly discussing changes in fragility with pressure. *Generally, all these computations point to variations of fragility between different polymers as arising from the relative efficiency of packing complex shaped molecules. In simpler terms, more deformable molecules fill space better than stiffer molecules, leading to stronger fluids that are less sensitive to the structural changes induced by temperature variation.* This simple physical picture of the structural origin of variations in fragility in glass-forming liquids is one of the principal conclusions of the present chapter.

Recent experiments indicate that glass formation can occur at *constant temperature* through an increase of pressure, providing a test for the entropy theory because a reliable theory should, however, entail *all* thermodynamic paths to glass formation within a unified framework. Specifically, an empirical pressure analogue of the VFTH equation has been advocated [175, 176] for the structural relaxation time $\tau(P, T = \text{constant})$ at a variable pressure. The pressure equation exhibits an apparent singularity at a critical pressure P_o, which is an analog to the Vogel temperature T_∞ of the original VFTH equation. Since it is not *a priori* obvious whether this relation for $\tau(P, T = \text{constant})$ can be derived within the entropy theory, we directly compute $\tau(P, T = \text{const})$. Consistency of the LCT entropy theory with the pressure-dependent relaxation time measurements is illustrated in Fig. 21, which indicates that the reduced configurational entropy $\delta s_c \equiv [s_c(T, P = 1\ \text{atm}) - s_c(T, P)]/s_c(T, P)$ at fixed T is inversely proportional to the reduced pressure $\delta P \equiv |P - P_o|/P$, where P_o is the Vogel pressure at which s_c vanishes or, equivalently, $\tau(P, T = \text{const})$ diverges. The agreement between our calculations in Fig. 21 and measurements is remarkable. Although the pressure analogue of the VFTH equation has been first discovered experimentally, we explain the origin of this relation and compute P_o in terms of the molecular parameters of our schematic model for glass formation.

The majority of our calculations are performed for models in which the side group length n in both F-F and F-S polymer chains is fixed ($n = 3$). To gauge the influence of n on glass formation and to check the consistency of the entropy theory with experiment, the glass transition temperature T_g is evaluated for the extended schematic model in which polymers are permitted to have side groups with variable lengths n. These calculations are conducted assuming that all characteristic parameters (ϵ, E_b, E_s, and M) remain constant, while n alone is varied. An increase of n leads to a sharp rise in T_g for F-S polymers and to a much weaker increase of T_g with n for F-F chains (see Fig. 19a). The predicted growth of T_g with n is not universal, however, as a sharp decrease of T_g with n is found for S-F polymers. Apparently, the manner in which free ends affect the glass transition temperature depends on the relative flexibility of the chain backbone and side groups, so that the classical "dangling chain end" argument [150] for the depression of T_g is not generally valid.

An important aspect of the present theoretical study is our ability to predict structural relaxation times in glass-forming polymer liquids based on a well defined molecular based statistical mechanical model. Specifically, the LCT for polymer melt glasses is combined with the AG model of dynamics in cooled liquids and an empirical relation between the high temperature activation energy $\Delta\mu$ and the crossover temperature T_I to provide a powerful framework for evaluating structural relaxation times τ without any adjustable parameters beyond those inherent in the schematic model of glass formation (ϵ, E_b, E_s). Illustrative examples of these calculations are presented in Figs. 20a and 20b, and the results are consistent with typical orders of magnitudes for experimentally determined $\tau(T_{mc}^{exp})$ and $\tau(T_g)$. Our calculations also reveal that the rescaling of the reduced temperature T_g/T, a variable related to breadth of glass formation (as measured by $T_I/(T_I - T_g)$), leads to the collapse of the ratio $\tau/\tau(T_I)$ for both fragile and strong polymer fluids into a universal curve, as found empirically by Rössler et al. [142].

The relation between the entropy and free volume description of glass formation is shown to produce similar pictures for the temperature dependence of τ at constant pressure for temperatures near T_g. Our main motivation for considering this alternative free volume description of glass-formation, however, is to determine the glass transition temperature through a Lindemann criterion. This Lindemann procedure, or something like it [defining T_g such that $\tau(T_g)$ is the constant value $\sim\mathcal{O}(10^2\text{–}10^3\ \text{s})$] is required since the entropy theory does not predict the existence of a thermodynamic signature at the glass transition temperature T_g. Our discussion of the relation between the excess free volume and the configurational entropy also relates s_cT to the Debye–Waller factor $\langle u^2\rangle$ for the amplitude of the thermal motions of particles (monomers) in the cooled liquids. Moreover, the Buchenau relation for τ as a function of $\langle u^2\rangle$ is deduced as an alternative representation of the AG equation (33). The characteristic temperatures of glass formation (T_0, T_g, T_I, T_A) all have counterparts in characteristic values for $\langle u^2\rangle$.

A direct quantitative comparison between AG theory and measurements requires the resolution of two issues. First, the excess entropy S_{exc} must be normalized by the molar volume. We suggest that the lack of this normalization is partly responsible for previous claims [15, 49] that AG theory breaks down for small molecule fluids. Second, the vibrational contribution to S_{exc}, which is absent in s, must be subtracted reliably. While the first correction can readily be introduced, the inclusion of the second correction requires further investigation [63, 240].

Another important limitation of our LCT-AG entropy theory of glass formation arises because the AG model implicitly focuses on large scale structural relaxation processes and cannot currently describe relaxation processes in the nonzero wavevector limit. This restriction to long wavelengths precludes treating many aspects of glass formation, such as the wavevector dependence of the structural relaxation time and the bifurcation of relaxation times. Thus, an important direction for the future extension of the entropy theory involves adding a square gradient contribution to the free energy (modeling the energetic cost of density fluctuations in the polymer melt) in order to describe the rate of structural relaxation at finite length scales.

A wide variety of theories have been proposed to describe glass formation, ranging from thermodynamic theories, such as the entropy and free volume models, to kinetic theories, such as mode coupling theory, that emphasize particle localization as the origin of structural arrest [7, 8, 10]. Each of these theories captures aspects of glass formation in real fluids, and it is thus natural to expect some interrelations between these alternative approaches to glass formation and the entropy theory. For example, Section X shows that the usual phenomenological free volume expression for the rate of structural relaxation of glass-forming fluids at low temperatures (the low temperature regime of glass formation) and constant pressure is recovered from the entropy theory. Our comparative analysis in Section XI between the entropy theory and other theories establishes that many conceptual and predictive characteristics are shared by these approaches. The correspondences are particularly clear in models of glass formation that are based on the concept of an activation energy that depends on thermodynamic properties. The discussion in Section XI begins with the recent approach of Schweizer and Saltzman [39, 40] (SS) who append an activation barrier to mode coupling theory to describe hopping processes in the fluid that augment the caging process emphasized in earlier versions of mode coupling theory. A rough comparison between this theory and the generalized entropy theory suggests that these theories exhibit a similar phenomenology in the high temperature regime of glass formation. Moreover, both SS and entropy theories display the Rössler scaling [142] when the crossover temperature T_x is treated as an adjustable parameter, and when T_x is replaced by T_I, respectively. The striking parallels between the SS and generalized entropy theories of glass formation suggest an approximate relation between the isothermal compressibility and the product of the temperature and the configurational entropy, a relation that should be examined futher in the future. A consideration of the relation between the configurational entropy and the isothermal compressibility, in turn, leads naturally to a connection of the entropy theory with the shoving model in which glass formation is driven by the sharply increasing elastic constants as the fluid is cooled. Lastly, the Buchenau relation [58] (consistent with the entropy theory) determines a link between thermodynamics (the configurational entropy) and particle localization, a viewpoint also emphasized by mode coupling theory. In particular, Eq. (64) relates the structural relaxation times and Debye–Waller factors and, thereby, provides yet another connection with free volume type models since the mean square amplitude of particle motion is determined by the excess free volume. Thus, the new entropy theory unifies these seemingly disparate concepts within a single framework that permits quantitative computations for the dependence of melt properties on monomer structure over temperature ranges in which the system can be equilibrated in the fluid state.

Although the entropy theory of glass formation provides many insights into the structural origin of fragility variations in polymer liquids and into the thermodynamic significance of the characteristic temperatures of glass formation, the theory is mute regarding the molecular scale dynamic processes responsible for many other aspects of glass formation. In Section XII, we grapple with the questions concerning the nature of the cooperatively rearranging regions (CRRs) in the AG theory. Assuming that such structures (and their immobile particle cluster counterparts) have a polymeric form, we speculate about sizes of these structures and how they lead to the pattern of complex fluid viscoelasticity that is observed experimentally.

Recent simulations [113, 114, 199, 200] and experiments [195, 196] for colloidal fluids both suggest that the CRRs take the form of string-like structures. This viewpoint clearly motivates comparing the thermodynamics of cooled polymer melts undergoing glass formation to the thermodynamics of polymeric clusters arising from equilibrium polymerization upon cooling (not necessarily taking the form of linear chains [206]). These two classes of complex fluids are indeed found to exhibit remarkable similarities in their thermodynamic behavior. In each case, the thermodynamic transition is accompanied by a drop of the configurational entropy to a low temperature residual value. Both are demarked by temperatures where this drop initiates (T_x in equilibrium polymerization [205]) and ends ($T_{\rm sat}$ in equilibrium polymerization [205]), and there is an inflection point in s_c (or in $s_c T$) at an intermediate temperature separating high and low temperature regimes of both these transitions. This resemblence strongly suggests that glass formation is itself a rounded transition as in equilibrium polymerization. The above view also accords in spirit (but not in details) with the frustration model of glass formation by Kivelson and co-workers [13], which likewise emphasizes the importance of frustration in packing as the control parameter for the extent of transition rounding [241]. As a further test of the correspondence between glass formation and equilibrium polymerization, we analyze the relation between s_c and the average mass of the CRRs (proportional to average chain length in the equilibrium polymerization model). Adam and Gibbs hypothesize that these quantities should be inveresly related to each other, and explicit calculations for two different models of equilibrium polymerizations (where z^* corresponds to the average degree of polymerization L) indicate that such a relation holds to a good approximation, thus providing another striking confirmation of the hypothesis that the CRRs are themselves both polymeric and dynamic by nature. The suggestion that liquids are built up from polymeric structures whose degree of polymerization changes with temperature has long been advocated for the description of cooled liquids [242].

The polymeric interpretation of excitation structures in glass-forming liquids also provides insights into the dynamical origins of variations of fragility in supercooled liquids. It is well known that chemical initiation and thermal activation ultimately regulate the extent of polymer growth in polymerization upon cooling and the sharpness (fragility) of the resulting transitions. It is reasonable to imagine that highly energetically unstable regions form in cooled liquids through thermal fluctuations and that these events serve as catalysts for collective motions [226]. In accordance with this analogy, the rate of change of the viscosity with temperature in living polymerization solutions can be tuned by varying the initiator concentration [224, 225].

The suggested correspondence between glass formation and equilibrium polymerization also provides a tentative framework for comprehending many nonequilibrium rheological properties of glass-forming materials. Physical aging can be naturally understood by the slow growth of the polymer chains into their equilibrium populations upon lowering temperature, as also found for equilibrium polymers. The viscosity and relaxation times grow as these structures form, and then the properties cease to evolve once equilibrium is achieved. Experimental measurements for equilibrium polymerization solutions become dependent on the scanning rate when the temperature is scanned too rapidly compared to the equilibration time. Similar nonequilibrium history effects are also characteristic of glass-forming liquids. Strong shear thinning is also prevalent in both glass-forming liquids and polymerizing liquids since shear tends to break down these structures, and aging naturally follows this "rejuvenating" process as these fluids recover their equilibrium structures. We hope that our increasing understanding of dynamic heterogeneity in equilibrium polymerization solutions can be translated into a corresponding theory of dynamic clustering in glass-forming liquids.

APPENDIX A: OBJECTIVE MEASURES OF FRAGILITY AND AN ORDER PARAMETER FOR GLASS FORMATION

Although the general concept of fragility as a measure of the strength of the temperature dependence of structural relaxation times and associated transport properties is unequivocal, some suggested specific measures of fragility, such as the apparent activation energy $\mathcal{E}$ (for structural relaxation) at T_g divided by T_g, seem to lack a fundamental significance [243]. In particular, these definitions depend on the range of the particle interactions since $\mathcal{E}$ depends on the cohesive energy, even at high temperatures where $\mathcal{E}(T \to \infty) \approx \Delta\mu$, and on molecular parameters. (This criticism would not apply if it is ultimately found that $\Delta\mu \propto T_g$, where the constant of proportionality is universal.) The sensitivity of T_g to cooling rate also raises questions about fundamental definitions of fragility based on this temperature. Estimates of the fragility parameter D in the VFTH equation for τ can vary widely depending on the temperature range chosen [244], so that this fragility definition can also be problematic. Ratios of the characteristic temperatures of glass formation seem to provide objective measures of the transition broadness and thus fragility, but the apparent lack of a fundamental thermodynamic significance of T_g engenders concerns about this type of fragility estimates as well. The combination of equilibrium polymerization theory, the generalized entropy theory of glass formation, and the identification of the strings with the CRRs of AG offers the prospect of a more fundamental definition of fragility.

Our goal lies in introducing a more fundamental definition for the strength of the temperature dependence of transport properties that is not subject to the criticism just mentioned and that equally applies to the high and low temperature regimes of glass formation. This extension is especially needed for the analysis of simulations since the low temperature regime is generally inaccessible in equilibrated simulations of glass-forming liquids.

The "differential fragility" χ_g of a glass-forming fluid at a temperature T may be defined through the derivative of the normalized activation energy for structural relaxation,

$$\chi_g = d[\mathcal{E}(T)/\mathcal{E}(T \to \infty)]/dT \tag{A1}$$

This quantity describes how much the temperature dependence of structural relaxation deviates from an Arrhenius behavior, which is the most basic concept embodied by the term "fragility." Within AG theory, this quantity exactly equals

$$\chi_{g,AG} = d[z^*(T)]/dT \tag{A2}$$

where z^* is the average number of particles in the cooperatively rearranging regions (CRRs). Furthermore, the identification of the CRRs with strings implies

$$\chi_{g,L}(strings \equiv CRRs) = d[L(T)]/dT \tag{A3}$$

It is apparent that χ_g corresponds to a kind of "susceptibility" associated with this type of rounded transition, and the term "glass susceptibility" might be a better term than "fragility." At any rate, these measures of glass fragility imply that the fluid transport properties are strong at high temperatures, and then the fragility gradually increases upon cooling below an onset temperature T_A for glass formation. The rate of change of χ_g becomes largest near T_g and then becomes smaller again at lower temperatures where a return to an Arrhenius behavior has often been reported [67, 173, 245].

Our definition of fragility or glass susceptibility should be supplemented by further information indicating the relative *location* of the fluid within this broad thermodynamic transition. Thus, we seek an objective measure of the relative breadth of this type of transition. The extent Φ of string polymerization provides a natural *order parameter for glass formation* (i.e., an "extent of glassiness") that quantifies the degree to which the glass-formation process has developed. In a previous paper [205], we showed that Φ can be approximately related to L in the $\mathcal{I}$ model of equilibrium polymerization by $\Phi \approx (L-1)/L$ or $L \approx 1/(1-\Phi)$, so that the ratio $(L-1)/L$ might also serve as an effective order parameter for glass formation. Both of these quantities vanish in the high temperature homogeneous fluid state and approach unity at low temperatures where the clustering process is finished. Consequently, either Φ or L can then serve as an objective measure of the extent of glassiness. A measure of the breadth of the polymerization or glass transition is obtained by considering the variance of L or Φ, or the absolute value of the derivative of Φ with respect to temperature, $|d\Phi(T)/dT|$. The half-width of these peaked functions provides an intrinsic measure of transition broadness.

These measures of transition susceptibility, extent of ordering, and transition breadth apply equally well to equilibrium polymerization and self-assembly as to glass formation. Thus, these measures offer a unified description of the characteristics of rounded transitions. The $\chi_{g,L}$ in Eq. (A3) is a susceptibility measure where L corresponds to the average cluster relative mass, the order parameter Φ describes the extent of conversion of particles to the cluster state, and the variance in L or Φ reflect the transition broadness [246].

Acknowledgments

This work is supported, in part, by NSF grant CHE 0416017. We are grateful to Arun Yethiraj for performing MD simulations for chains of our schematic model of glass formation and thank Alexei Sokolov and Ken Schweizer for helpful discussions and comments.

References

1. W. Kurz and D. J. Fisher, *Fundamentals of Solidification*, (4th ed., Trans Tech Publ Ltd., Switzerland, 1998.
2. C. A. Angell, *Science* **267**, 1924 (1995).
3. C. A. Angell, *J. Non-Cryst. Solids* **131–133**, 13 (1991).
4. L. -M. Martinez and C. A. Angell, *Nature* **410** 663 (2001).
5. The simple interpretation of fragility as a measure of the rate at which η changes with temperature only applies to substances that have similar activation energies $\Delta\mu$.
6. J. L. Green, K. Ito, K. Xu, and C. A. Angell, *J. Phys. Chem. B* **103**, 3991 (1999).
7. M. D. Ediger, C. A. Angell, and S. R. Nagel, *J. Phys. Chem.* **100**, 13200 (1996).
8. G. B. McKenna, in *Comprehensive Polymer Science*, Vol. 2 (C. Booth and C. Price, eds.), Pergamon, Oxford, UK, 1989, p. 311.
9. K. L. Ngai, *J. Non-Crvst. Solids* **275**, 7 (2000).
10. P. G. Debenedetti and F. H. Stillineer. *Nature* 410, 259 (2001).
11. B. Zhang, D. Q. Zhao, M. X. Pan, W. H. Wang, and A. L. Greer, *Phys. Rev. Lett.* **94**, 205502 (2005).
12. C. Dodd and H. P. Mi, *Proc. Phys. Soc. A* **62**, 464 (1949); N. N. Greenwood and R. L. Martin, *Proc. R. Soc.* **215**, 46 (1952).
13. D. Kivelson, S. A. Kivelson, X. Zhao, Z. Nussinov, and G. Tarjus, *Physica A* **219**, 27 (1995); D. Kivelson, G. Tarjus, X. Zhao, and S. A. Kivelson, *Phys. Rev. E* **53**, 751 (1996).
14. E. W. Fisher, *Physica A* **201**, 183 (1993).
15. K. L. Ngai, *J. Chem. Phys.* **111**, 3639 (1999); *J. Phys. Chem. B* **103**, 5895 (1999).
16. F. Fujara, B. Geil, H. Sillescu, and G. Fleischer, *Z. Phys. B* **88**, 195 (1992).
17. M. T. Cicerone, F. R. Blackburn, and M. D. Ediger, *J. Chem. Phys.* **102**, 471 (1995).
18. M. D. Ediger, *Annu. Rev. Phys. Chem.* **51**, 99 (2000).
19. W. Götze and L. Sjögren, *Rep. Prog. Phys.* **55**, 241 (1992).
20. A. Schönhals, F. Kremer, A. Hoffmann, E. W. Fisher, and E. Schlosser, *Phys. Rev. Lett.* **70**, 3459 (1993).
21. Y. Brumer and D. R. Reichman, *Phys. Rev. E* **69**, 041202 (2004).
22. J. Dudowicz, K. F. Freed, and J. Douglas, *J. Chem. Phys.* **123**, 111102 (2005).
23. F. E. Simon, *Naturwiss enschaften* **9**, 244 (1930).
24. W. Kauzmann, *Chem. Rev.* **43**, 219 (1948).
25. C. M. Roland, S. Capaccioli, M. Lacchesi, and R. Casalini, *J. Chem. Phys.* **120**, 10640 (2004).
26. D. Huang, D. M. Colucci, and G. B. McKenna, *J. Chem. Phys.* **116**, 3925 (2002).
27. E. Leutheusser, *Phys. Rev. A* **29**, 2765 (1984); U. Bengtzeluis, W. Götze, and A. Sjolander, *J. Phys. Cond. Matt.* **17**, 5915 (1984).
28. W. Götze, *J. Phys. Cond. Matt.* **11**, Al (1999); *Physica A* **235**, 369 (1997).
29. K. Kawasaki, *Phase Transitions and Critical Phenomena* **5a**, 165 (1976).
30. J. G. Kirkwood and E. Monroe, *J. Chem. Phys.* **9**, 314 (1941).
31. J. D. Weeks, S. A. Rice, and J. J. Kozak, *J. Chem. Phys.* **52**, 2416 (1976); I. S. Chang and J. J. Kozak, *J. Math. Phys.* **14**, 632 (1973).
32. H. J. Ravache and R. F. Kayser, *J. Chem. Phys.* **68**, 3632 (1978).

33. W. Klein and N. Grewe, *J. Chem. Phys.* **72**, 5456 (1980).
34. W. Klein, H. Gould, R. A. Ramos, I. Clejan, and A. I. Mel'cuk, *Physica A*, **205**, 738 (1994); N. Grewe and W. Klein, *J. Math. Phys.* **18**, 1735 (1979); J. W. Haus and P. H. E. Meijer, *Phys. Rev. A* **14**, 2285 (1976).
35. G. Johnson, A. I. Mel'cuk, H. Gould, W. Klein, and R. D. Mountain, *Phys. Rev. E* **57**, 5707 (1998); A. I. Mel'cuk, R. A. Ramos, H. Gould, W. Klein, and R. D. Mountain, *Phys. Rev. Lett.* **75**, 2522 (1995).
36. W. Klein and A. D. J. Haymet, *Phys. Rev. B* **30**, 1387 (1984).
37. L. Leibler, *Macromolecules* **13**, 1602 (1980).
38. W. Klein, personal communication.
39. K. S. Schweizer and E. J. Saltzman, *J. Chem. Phys.* **119**, 1181 (2003); *ibid.* **121**, 1984 (2003); Sjögren [Z. Phys. B **79**, 5 (1980)] has suggested the scaling of the apparent activation energy with inverse compressibility discussed in our Eqs. (56)–(58).
40. E. J. Saltzman and K. S. Schweizer, *J. Chem. Phys.* **121**, 2001 (2004).
41. T. V. Ramakrishnan and M. Yussouff, *Phys. Rev. B* **19**, 2775 (1979); A. D. J. Haymet, *J. Chem. Phys.* **89**, 887 (1985).
42. X. Xia and P. G. Wolynes, *Proc. Natl. Acad. Sci. USA* **97**, 2990 (2000).
43. R. W. Hall and P. G. Wolynes, *J. Chem. Phys.* **86**, 2943 (1987).
44. C. Kaur and S. P. Das, *Phys. Rev. Lett.* **86**, 2062 (2001).
45. J. -P. Bouchaud and G. Biroli, *J. Chem. Phys.* **121**, 7347 (2004).
46. The theory formally can be extrapolated to the limit of monomers and thus can be applied to describe nonpolymeric glass-forming liquids.
47. J. H. Gibbs and E. A. DiMarzio, *J. Chem. Phys.* **28**, 373 (1958).
48. G. Adam and J. H. Gibbs, *J. Polym. Sci.* **40**, 121 (1959); *J. Chem. Phys.* **43**, 139 (1965).
49. R. Richert and C. A. Angell, *J. Chem. Phys.* **108**, 9016 (1998).
50. K. F. Freed, *J. Chem. Phys.* **119**, 5730 (2003).
51. K. W. Foreman and K. F. Freed, *Adv. Chem. Phys.* **103**, 335 (1998).
52. J. Dudowicz, K, F. Freed, and J. F. Douglas, *J. Chem. Phys.* **124**, 064901 (2006).
53. J. Dudowicz, K. F. Freed, and J. Douglas, *J. Phys. Chem. B* **109**, 21285 (2005).
54. J. Dudowicz, K. F. Freed, and J. Douglas, *J. Phys. Chem. B* **109**, 21350 (2005).
55. K. Binder, J. Baschnagel, and W. Paul, *Prog. Polym. Sci.* **38**, 115 (2003).
56. F. A. Lindemann, *Z. Phys.* **11**, 609 (1910); J. J. Gilvarry, *Phys. Rev.* **102**, 308 (1956).
57. J. C Dyre, N. B. Olsen, and T. Christensen, *Phys. Rev. B* **53**, 2171 (1996); J. C. Dyre, *J. Non-Crvst. Solids* **235–237**, 142 (1998), J. C. Dyre and N. B. Olsen, *Phys. Rev. E* **69**, 042501 (2004); S. V. Nemilov, *Russ. J. Phys. Chem.* **42**, 726 (1968); F. Bueche, *J. Chem. Phys.* **30**, 748 (1959); A. Tobolsky, R. E. Powell, and H. Eyring, in *Frontiers in Chemistry*, Vol. 1 (R. E. Burke and O. Grummit, eds.), Interscience, New York, 1943, P. 125.
58. U. Buchenau and R. Zorn, *Europhys. Lett.* **18**, 523 (1992).
59. K. H. Meyers, *Z. Phys. Chem. B (Liepzig)* **44**, 383 (1939); P. J. Flory, *J. Chem. Phys.* **10**, 51 (1942); M. L. Huggins, *J. Phys. Chem.* **46**, 151 (1942).
60. J. H. Gibbs, *J. Chem. Phys.* **25**, 185 (1956); E. A. DiMarzio and J. H. Gibbs, *J. Chem. Phys.* **28**, 807 (1958); E. A. DiMarzio, *Ann. NY Acad. Sci.* **371**, 1 (1981).
61. E. A. DiMarzio and A. J. Yang, *J. Res, NIST* **102**, 135 (1997).
62. E. A. DiMarzio, *Comp. Mater Sci.* **4**, 317 (1995).

63. G. P. Johari. *J. Chem. Phys.* **112**, 8958 (2000).
64. A. Scala, F. W. Starr, E. La Nave, F. Sciortino, and H. E. Stanley, *Nature* **406**, 166
65. S. Mossa, E. La Nave, H. E. Stanley, C. Donati, F. Sciortino, and P. Tartaglia, *Phys. Rev. E* **65**, 041205 (2002).
66. S. Corezzi, D. Fioretto, and P. Rolla, *Nature* **420**, 653 (2002).
67. I. Saika-Voivod, P. H. Poole, and F. Sciortino, *Nature*, **412**, 514 (2001).
68. I. Saika-Voivod, F. Sciortino, and P. H. Poole, *Phys. Rev. E* **99**, 041503 (2004).
69. H.-J. Oels, and G. Rehage, *Macromolecules* **10**, 1036 (1977).
70. P. D. Gujrati and M. Goldstein, *J. Chem. Phys.* **74**, 2596 (1981).
71. M. Gordon, K. Kapadia, and A. Malakis, *J. Phys. A* **9**, 751 (1976).
72. M. Mèzard and G. Parisi, *J. Chem. Phys.* **111**, 1076 (1999); *Phys. Rev. Lett.* **82**, 747 (1999); *J. Phys. C* **12**, 6655 (2000).
73. M. Wolfgardt, J. Baschnagel, W. Paul, and K. Binder, *Phys. Rev. E* **54**, 1535 (1996).
74. A. I. Milchev, *C. R. Bulg. Sci.* **36**, 139 (1983).
75. M. L. Williams, R. F. Landel, and J. D. Ferry, *J. Am. Chem. Soc.* **77**, 3701 (1955); J. D. Ferry, *Viscoelastic Properties of Polymers, 3rd ed., John Wiley & Sons, Hoboken, NJ, 1980; A. Schönhals, F. Kremer, A. Hofmann, E. W. Fischer, and E. Schlosser, Phys. Rev. Lett.* **70**, 3459 (1993).
76. There are instances, especially in polymeric fluids, in which T_K is quite distinct from T_0, but both these temperatures lie generally below T_g. Recent studies [53, 106, 107] also reveal that the temperature at which $S_{exc}^{(mol)}$ extrapolates to zero is not equivalent to the Vogel temperature T_∞ for some polymer fluids. This finding implies the inequivalence of the configurational entropy and excess fluid entropy even when these two quantities are normalized in the same way.
77. K. L. Ngai and C. M. Roland, *Macromolecules* **26**, 6824 (1993).
78. D. M. Colucci and G. B. McKenna, *Mater Res. Soc. Symp. Proc.* **455, 171** (1997).
79. P. G. Santangelo and C. M. Roland, *Macromolecules* **31**, 4581 (1998).
80. The symbol z^* is used in the literature for the ratio s_c^*/s_c and should not be confused with the lattice coordination number z.
81. J. F. Kincaid, H. Eyring, and A. E. Stearn, *Chem. Rev.* **28**, 301 (1941). See also R. M. Barker, *Trans. Faraday Soc.* **39**, 48 (1943).
82. V. A. Bershtein, V. M. Egorov, L. M. Egorova, and V. A. Ryzhov, *Thermochim. Acta* **238**, 41 (1994).
83. R. Zwanzig and A. K. Harrison, *J. Chem. Phys.* **83**, 5861 (1985).
84. D. Tabor, *Philos. Mag. A* **57**, 217 (1988).
85. S. Corezzi, D. Fioretto, and J. M. Kenny, *Phys. Rev. Lett.* **94**, 065702 (2005).
86. Considering the permutational collective motion in a fluid, Feynman (Phys. Rev. **91**, 1291 (1953)) argues that the probability of an *n*th particle collective displacement is on the order of the *n*th power of the probability of an elementary binary displacement event. This argument is equivalent to the first AG hypothesis if the CRRs of cooled classical fluids are identified with the string-like particle permutational motions that Feynman discusses for liquid He near its superfluid transition. Feynman treates the superfluid transition in He as a kind of equilibrium polymerization transition, and in a separate paper [246], we analyze glass-forming liquids from a similar phenomenological perspective.
87. U. Mohanty, I. Oppenheim, and C. H. Taubes, *Science* **266**, 425 (1994).
88. A. B. Bestul and S. S. Chang, *J. Chem. Phys.* **40**, 3731 (1964).

89. S. Takahara, O. Yamamuro, and H. Suga, *J. Non-Cryst. Solids* **171**, 259 (1994).
90. J. H. Magill, *J. Chem. Phys.* **47**, 2802 (1967).
91. C. A. Angell, *J. Res. NIST* **102**, 171 (1997).
92. S. Sastry, *Nature* **409**, 164 (2001).
93. V. Lubchenko and P. Wolynes, *J. Chem. Phys.* **119**, 9088 (2003).
94. F. H. Stillinger and T. A. Weber, *Phys. Rev. A* **25**, 978 (1982).
95. The configurational entropy obtained from the simulations is calculated by sampling the number of basins in the potential energy landscape and by *identifying* the minima of these basins with distinguishable fluid configurations that are used in defining the thermodynamic entropy (see Refs. 64–68).
96. K. F. Freed and J. Dudowicz, *Adv. Polym. Sci.* **183**, 63 (2005).
97. J. Dudowicz and K. F. Freed, *Macromolecules* **28**, 6625 (1995); ibid. **30** 5506 (1997); H. Frielinghaus, D. Schwahn, J. Dudowicz, and K. F. Freed, *J. Chem. Phys.* **114**, 5016 (2001).
98. J. Dudowicz, M. S. Freed, and K. F. Freed, *Macromolecules* **24**, 5096 (1991); J. Dudowicz and K. F. Freed, *Macromolecules* **24**, 5112 (1991); K. F. Freed and J. Dudowicz, *Theor. Chim. Acta* **82**, 357 (1992); J. Dudowicz and K. F. Freed, *J. Chem. Phys.* **96**,1644 (1992); *ibid.* **96**, 9147 (1992); J. Dudowicz, K. F. Freed, and J. F. Douglas, *Macromolecules* **28**, 2276 (1995); K. F. Freed and J. Dudowicz, *Trends in Polym. Sci.* **3**, 248 (1995); K. F. Freed and J. Dudowicz, *Macromolecules* **29**, 625 (1996); J. Dudowicz and K. F. Freed, *Macromolecules* **29**, 7826 (1996); *ibid.* **29**, 8960 (1996); K. F. Freed, J. Dudowicz, and K. W. Foreman, *J. Chem. Phys.* **108**, 7881 (1998); J. Dudowicz and K. F. Freed, *Macromolecules* **31**, 5094 (1998); K. F. Freed and J. Dudowicz, *Macromolecules* **31**, 6681 (1998); C. Delfolie, L. C. Dickinson, K. F. Freed, J. Dudowicz, and W. J. MacKnight, *Macromolecules* **32**, 7781 (1999); J. Dudowicz and K. F. Freed, *Macromolecules* **33**, 3467 (2000); *ibid.* **33**, 5292 (2000); *ibid.* **33**, 9777 (2000); J. Dudowicz, K. F. Freed, and J. F. Douglas, *Phys. Rev. Lett*, **88**, 095503 (2002); J. Dudowicz, K. F. Freed, and J. F. Douglas, *J. Chem. Phys.* **116**, 9983 (2002).
99. G. Floudas and P. Štepánek, *Macromolecules* **31**, 6951 (1998).
100. A. Yethiraj, personal communication.
101. K. W. Foreman and K. F. Freed, *Macromolecules* **30**, 7279 (1997).
102. V. N. Novikov and A. P. Sokolov, *Phys. Rev. E* **67**, 031507 (2003); I. M. Hodge, *J. Non-Cryst. Solids* **202**, 164 (1996).
103. H. Vogel, *Phys. Z.* **22**, 645 (1921); G. S. Fulcher, *J. Am. Ceram. Soc.* **8**, 339 (1925); G. Tammann and W. Hesse, *An. Allg. Chem.* **156**, 245 (1926).
104. C. A. Angell, *Pure Appl. Chem.* **63**, 1387 (1991).
105. S.-J. Kim and T. E. Karis, *J. Mater Res.* **10**, 2128 (1995).
106. D. Cangialosi, A. Alegria, and J. Colmenero, *Europhys. Lett.* **70**, 614 (2005).
107. D. Prevosto, M. Lucchesi, S. Capaccioli, R. Casalini, and P. Rolla, *Phys. Rev. B* **67**, 174202 (2003).
108. S. Kamath, R. H. Colby, S. K. Kumar, and J. Baschnagel, *J. Chem. Phys.* **116**, 865 (2002).
109. Actually, simulation data [108] for $s_{c,L}(T)$ exhibit a shallow maximum.
110. J. S. Langer, *Phys. Rev. E* **70**, 041502 (2004). See also M. L. Falk, J. S. Langer, and L. Pechenik, *Phys. Rev. E* **70**, 011507 (2004).
111. S. Matsuoka, *Relaxation in Polymers* Hanser Publishers, New York, 1992. T_A is sometimes denoted as T^*.

112. At temperatures much higher than T_A, the polymer fluid formally undergoes a fluid–gas transition where the isothermal compressibility κ_T diverges, but this high temperature regime is normally inaccessible in polymer systems because of thermal decomposition.

113. C. Donati, J. F. Douglas, W. Kob, S. J. Plimpton, P. H. Poole, and S. C. Glotzer, *Phys. Rev. Lett.* **80**, 2338 (1998). There are interesting earlier suggestions of string-like or tunnel-like motion in cooled colloidal fluids and shaken granular fluids. See A. Rahman, *J. Chem. Phys.* **45**, 2585 (1966); R. Zwanzig and M. Bishop, *J. Chem. Phys.* **60**, 295 (1974).

114. C. Bennemann, C. Donati, J. Baschnagel, and S. C. Glotzer, *Nature* **399**, 246 (1999).

115. N. Giovambattista, S. V. Buldyrev, F. W. Starr, and H. E. Stanley, *Phys, Rev. Lett.* **90**, 085506 (2003).

116. N. Giovambattista, S. V. Buldyrev, H. E. Stanley, and F. W. Starr, *Phys. Rev. E* **72**, 011202 (2005). The strings in this investigation of the dynamics of water are actually mobile particle clusters, as described in Refs. 113 and 115.

117. E. V. Russell and N. E. Israeloff, *Nature* **408**, 695 (2000); E. V. Russell, N. E. Israeloff, L. E. Walther, and H. A. Gomariz, *Phys. Rev. Lett* **81**, 1461 (1998).

118. The temperature dependence of τ in the high temperature regime ($T > T_T$) is often modeled by an alternative VFTH equation.

119. M. Beiner, H. Huth, and K. Schröter, *J. Non-Cryst. Solids* **279**, 126 (2001).

120. J. Bascle, T. Garel, and H. Orland, *J. Phys. A* **25**, L-1323 (1992).

121. S. M. Aharony, *J. Macromol. Sci. Phys.* **B9**, 699 (1974); *J. Polym. Sci. C* **42**, 795 (1973). The onset temperature T_A for glass formation is denoted in these references as T_R.

122. S. Sastry, P. G. Debedenetti, and F. H. Stillinger, *Nature* **393**, 554 (1998).

123. Y. Brumer and D. R. Reichman, *Phys. Rev. E* **69**, 041202 (2004). This paper indicates that the theoretically determined mode coupling temperature T_{mc} is near T_A, while fits of the mode coupling theory parameters to experimental data, as well as the results of our calculations, suggest that the experimentally estimated T_{mc}^{exp} actually or nearly coincides with T_I.

124. F. W. Starr and J. F. Douglas, unpublished results.

125. J. Colmenero, A. Arbe, G. Coddens, B. Frick, C. Mijangos, and H. Reinecke, *Phys. Rev. Lett.* **78**, 1928 (1997).

126. M. Grimsditch and N. Rivier, *Appl. Phys. Lett.* **58**, 2345 (1999).

127. O. Yamamuro, I. Tsukushi, A. Lindqvist, S. Takahara, M. Ishikawa, and T. Matsuo, *J. Phys. Chem. B* **102**, 1605 (1998).

128. F. H. Stillinger, *J. Chem. Phys.* **89**, 6461 (1988).

129. J. H. Bilgram, *Phys. Rep.* **153**, 1 (1987); H. Löwen, *Phys. Rep.* **237**, 249 (1994).

130. R. A. LaViolette and F. H. Stillinger, *J. Chem. Phys.* **83**, 4079 (1985).

131. Y. Ding, A. Kisliuk, and A. P. Sokolov, *Macromolecules* **37**, 161 (2004).

132. X. Lu and B. Jiang, *Polymer* **32**, 471 (1991); A. A. Miller, *J. Polym. Sci. A* **6**, 249 (1968).

133. V. N. Novikov and A. P. Sokolov, *Nature* **431**, 961 (2004). Actually, they specifically claim that a closely related quantity to fragility, the steepness index m, is proportional to T_g.

134. A. Voronel, E. Veliyulin, T. Grande, and H. A. Oye, *J. Phys. Cond. Matt.* **9**, L247 (1997); E. Veliyulin, A. Voronel, and H. A. Oye, *J. Phys. Cond. Matt.* **7**, 4821 (1995). Voronel and co-workers find that $\Delta\mu$ equals 5.9 ± 0.1 times a temperature "closely related" to T_{mc}^{exp} that we identify with T_I. This result has been obtained for numerous salts. Note that it is difficult to estimate $\Delta\mu$ for polymers at high temperatures due to their tendency to thermal degradation at temperatures higher than T_A.

135. Since polymer materials are subject to thermal degradation at high temperatures, experimental estimates of the activation energy are often made at moderate temperatures ($T \simeq T_{\rm I}$) where $\mathcal{E} = \Delta\mu$ is not a good approximation. Since z^* can be as large as 2 or 3 in this temperature range, the fitted constant of proportionality between the apparent activation energy and $T_{\rm mc}^{\rm exp}$ (or $T_{\rm I}$) can be larger for polymer fluids than indicated by Eq. (42).

136. L. Bottezzati and A. L. Greer, *Acta Metall.* **37**, 1791 (1989); Y. Hayashiuchi, T. Hagihora, and T. Okada, *Physica* **115B**, 67 (1982). A large body of data for metals and metal alloys indicate that $\Delta\mu$ is approximately proportional to the melting temperature $T_{\rm m}$. In view of the approximation $T_{\rm m} \simeq T_{\rm A}$ discussed in the text, this phenomenology also suggests a correlation between $\Delta\mu$ and $T_{\rm A}$.

137. A. P. Sokolov, personal communication.

138. Y. Ding, V. N. Novikov, A. P. Sokolov, A. Cailliaux, C. Dalle-Ferrier, and C. Alba-Simionesco, *Macromolecules* **37**, 9264 (2004).

139. D. J. Plazek and V. M. O'Rourke, *J. Polym. Sci.* **A-2**, 209 (1971).

140. K. L. Ngai and D. J. Plazek, *Physical Properties of Polymers* (J. Mark, ed.), American Institute of Physics, Woodbury, NY, 1996, Chap. 25.

141. D. J. Plazek, I. -C. Chay, K. L. Ngai, and C. M. Roland, *Macromolecules* **28**, 6432 (1995); D. J. Plazek, X. D. Zheng, and K. L. Ngai, *Macromolecules* **25**, 4920 (1992).

142. E. Rössler, K. -U. Hess, and V. N. Novikov, *J. Non-Cryst. Solids* **223**, 207 (1998).

143. A. P. Sokolov, *Science* **273**, 1675 (1996); A. P. Sokolov, A. Kisliuk, D. Quitmann, A. Kudlik, and E. Rössler, *J. Non-Cryst. Solids* **172–174**, 138 (1994).

144. R. G. Beaman, *J. T. Polym. Sci.* **9**, 470 (1952).

145. S. Sakka and J. D. Mackenzie, *J. Non-Cryst. Solids* **6**, 145 (1971).

146. R. F. Fedors, *J. Polvm. Sci.* **17**, 719 (1979); R. F. Boyer, Rubber Chem. Technol. **36**, 1303 (1963).

147. L. M. Torell, L. Börjesson, and A. P. Sokolov, *Transp. Theory Stat. Phys.* **24**, 1097 (1995).

148. L. Santen and W. Krauth, *Nature* **405**, 550 (2000); S. Torquato, *Nature* **405**, 521 (2000).

149. No thermodynamic signature at $T_{\rm g}$ is evident in specific heat data at equilibrium, but a peak is observed under nonequilibrium conditions and is often taken as the *definition* of the glass transition. Unfortunately, this nonequilibrium peak cannot be addressed within the LCT of glass formation. We strictly avoid a discussion of the specific heat, given the complications of interpreting these data for polymer materials and the omission of the important vibrational component in the LCT treatment.

150. T. G. Fox and S. Loshaek, *J. Polym. Sci.* **15**, 371 (1955).

151. R. Simha and C. E. Weil, *J. Macromol. Phys. Sci. B* **4**, 215 (1970).

152. F. W. Starr, S. Sastry, J. F. Douglas, and S. C. Glotzer, *Phys. Rev. Lett.* **89**, 125501 (2002).

153. D. A. Young and B. J. Alder, *J. Chem. Phys.* **60**, 1254 (1974); R. Ohnesorge, H. Löwen, and H. Wagner, *Europhys. Lett.* **22**, 245 (1993).

154. C. A. Murray and D. G. Grier, *Annu. Rev. Phys. Chem.* **47**, 421–462 (1996).

155. R. Zwanzig and R. D. Mountain, *J. Chem. Phys.* **43**, 4464 (1965).

156. D. J. Plazek and J. H. Magill, *J. Chem. Phys.* **49**, 3678 (1968).

157. U. Buchenau and A. Wischnewski, *Phys. Rev. B* **70**, 092201 (2004).

158. G. T. Dee, T. Ougizawa, and D. J. Walsh, *Polymer* **33**, 3462 (1992).

159. K. Ueberreiter and G. Kanig, *J. Colloid Sci.* **7**, 569 (1953).

160. T. G. Fox and P. J. Flory, *J. Appl. Phys.* **21**, 581 (1950).

161. I. C. Sanchez, *J. Chem. Phys.* **79**, 405 (1983).

162. Q. S. Bhatia, J.-K. Chen, J. T. Koberstein, J. E. Sohn, and J. A. Emerson, *J. Colloid Interface Sci.* **106**, 353 (1985).

163. G. Dlubek, J. Pionteck, and D. Kilburn, *Macromol. Chem. Phys.* **205**, 500 (2004).

164. C. M. Stafford, C. Harrison, K. L. Beers, A. Karim, E. J. Amis, M. R. Vanlandingham, H. Kim, W. Volksen, R. D. Miller, and E. Simonyi, *Nature Mat.* **3**, 545 (2004).

165. D. O. Miles, *J. Appl. Phys.* **33**, 1422 (1962).

166. J. R. Lizotte, B. M. Erwin, R. H. Colby, and T. E. Long, *J. Polym. Sci. A.* **40**, 583 (2002).

167. B. M. Erwin, personal communication.

168. E. Donth, *J. Polym. Sci. B* **34**, 2881 (1996).

169. S. Rogers and L. Mandelkern, *J. Phys. Chem.* **61**, 985 (1957).

170. M. Wind, R. Graf, A. Heuer, and H. W. Spiess, *Phys. Rev. Lett.* **91**, 155702 (2003).

171. W. Kob and H. C. Anderson, *Phys. Rev. Lett.* **73**, 1376 (1994); F. W. Starr, F. Sciortino, and H. Stanley, *Phys. Rev. E* **60**, 6757 (1999).

172. J. Wuttke, M. Kiebel, E. Bartsch, F. Fujara, W. Petry, and H. Silescu, *Z. Phys. B* **91**, 357 (1993); F. X. Prielmeier, E. W. Lang, R. J. Sppedy and H.-D. Lündemann, *Phys. Rev. Lett.* **59**, 1128 (1987); P. Lunkenheimer, A. Pimenov, and A. Loidl, *Phys. Rev. Lett.* **78**, 2995 (1997); A. Meyer, J. Wuttke, W. Petry, O. G. Randl, and H. Schober, *Phys. Rev. Lett.* **80**, 4454 (1998).

173. P. A. O'Connell and G. B. McKenna, *J. Chem. Phys.* **110**, 11054 (1999).

174. N. Menon and S. R. Nagel, *Phys. Rev. Lett.* **74**, 1230 (1995).

175. M. Paluch, A. Patkowski, and E. W. Fischer, *Phys. Rev. Lett.* **85**, 2140 (2000).

176. K. Mpoukouvalas and G. Floudas, *Phys. Rev. E* **68**, 031801 (2003).

177. The calculated LCT configuration entropy s_c^*/k_B for high molar mass F-S polymer fluids at pressures of $P = 1$ atm and $P = 240$ atm equals 0.1933 and 0.2147, respectively, and can be regarded, to a first approximation, as pressure independent.

178. K. Schmidt-Rohr and H. W. Spiess, *Phys. Rev. Lett.* **66**, 3020 (1991).

179. The free volume δv treatment of glass formation can be extended to a description that encompasses constant volume conditions by considering a generalized reduced compressibility factor, $\delta Z = |PV - (PV)_0|/PV$, which reduces to δv and δP under constant pressure and constant volume conditions, respectively. The o subscript indicates that the product of the pressure P and the total volume V is be evaluated under the constraint of vanishing s_c.

180. C. M. Roland, S. Hensel-Bielowka, M. Paluch, and P. Casalini, *Rep. Prog. Phys.* **68**, 1405 (2005).

181. C. M. Roland and P. Casalini, *Phys. Rev. B* **71**, 014210 (2005).

182. M. Grousson, G. Tarjus, and P. Viot, *Phys. Rev. E* **65**, 065103 (2002).

183. F. Bueche, *J. Chem. Phys.* **21**, 1850 (1953).

184. J. M. Gordon, *J. Phys. Chem.* **82**, 963 (1973).

185. This approximation is not explicitly stated in Ref. 39.

186. The expression (59) reasonably describes data for the isothermal compressibility of many polymers over a wide temperature range if A and B are treated [39,40] as adjustable parameters.

187. R. Zwanzig and R. D. Mountain, *J. Chem. Phys.* **43**, 4464 (1965) show that the modulus G_∞ and the isothermal compressibility κ_T are determined by similar integrals containing the pair correlation function and the interparticle potential for simple Lennard-Jones fluids. The adiabatic (zero frequency) bulk modulus K_0 equals $-V(\partial P/\partial V)|_s$, which clearly is a kind

of reciprocal compressibility, and K_∞ is the high frequency analogue of K_0. For a nearly incompressible, solid-like medium at low T, we have $K_\infty \sim K_0 \sim G_\infty$ and $\kappa_T \sim \kappa_0$, leading to the rough approximation $G_\infty \sim 1/\kappa_T$. Zwanzig and Mountain also note that Cauchy's identity for isotropic solids composed of molecules interacting with two-body potentials implies that the ratio of the bulk modulus to shear modulus equals 5/3. The latter relation corresponds to a Poisson ratio of $\nu = 1/4$, which is a typical value for polymer glasses.

188. J. H. van Zanten and K. P. Rufener, *Phys. Rev. E* **62**, 5389 (2000).
189. T. Kamaya, T. Tsukushi, K. Kaji, J. Bartos, and J. Kristiak, *Phys. Rev. E* **60**, 1906 (1999).
190. S. Magazù, G. Maisano, and F. Migliardo, *J. Chem. Phys.* **121**, 8911 (2004); S. Magazù, G. Maisano, F. Migliardo, and C. Mondelli, *J. Phys. Chem. B*, **108**, 13580 (2004).
191. U. Tracht, M. Wilhelm, A. Heuer, H. Feng, K. Schmidt-Rohr, and H. W. Spiess, *Phys. Rev. Lett.* **81**, 2727 (1998).
192. I. Chang, F. Fujara, B. Geil, G. Heuberger, T. Mangel, and H. Sillescu, *J. Non-Cryst. Solids* **172–174**, 248 (1994); I. Chang and H. Sillescu, *J. Phys. Chem. B* **101**, 8794 (1997).
193. M. T. Cicerone and M. D. Ediger, *J. Chem. Phys.* **103**, 5684 (1995); *ibid.* **104**, 7210 (1996).
194. S. A. Reinsberg, X. H. Qui, M. Wilhelm, H. W. Spiess, and M. D. Ediger, *J. Chem. Phys.* **114**, 7299 (2001).
195. A. H. Marcus, J. Schofield, and S. A. Rice, *Phys. Rev. E.* **60**, 5725 (1999); B. Cui, B. Lin, and S. A. Rice, *J. Chem. Phys.* **114**, 9142 (2001).
196. E. R. Weeks, J. C. Crocker, A. C. Levitt, A. Schofield, and D. A. Weitz, *Science* **287**, 627 (2000).
197. W. K. Kegel and A. van Blaaderen, *Science.* **287**, 290 (2000).
198. D. Thirumalai and R. D. Mountain, *Phys. Rev. E* **47**, 479 (1993).
199. W. Kob, C. Donati, S. J. Plimpton, P. H. Poole, and S. C. Glotzer, *Phys. Rev. Lett* **79**, 2827 (1997).
200. C. Donati, S. C. Glotzer, P. H. Poole, W. Kob, and S. J. Plimpton, *Phys. Rev.E* **60**, 3107 (1999).
201. M. Dzugutov, S. I. Simdyankin, and F. H. M. Zetterling, *Phys. Rev. Lett.* **89**, 195701 (2002).
202. Y. Hiwatari and T. Muranaka, *J. Non-Cryst. Solids* **235–237**, 19 (1998); H. Miyagawa, Y. Hiwatari, Z. Bernu, and J. P. Hansen, *J. Chem. Phys.* **88**, 3879 (1988).
203. D. N. Perera and P. Harrowell, *J. Non-Cryst. Solids* **235–237**, 314 (1998); X. Wahnström, *Phys. Rev. A* **44**, 3752 (1991).
204. R. Yamamoto and A. Onuki, *Phys. Rev. E* **58**, 3515 (1998).
205. J. Dudowicz, K. F. Freed, and J. F. Douglas, *J. Chem. Phys.* **111**, 7116 (1999).
206. J. F. Douglas and J. B. Hubbard, *Macromolecules* **24**, 3163 (1991).
207. L. C. E. Struik, *Physical Aging in Amorphous Polymers and Other Materials*, Elsevier, Amsterdam, 1978.
208. A. Lee and G. B. McKenna, *Polym. Eng. Ska.* **30**, 431 (1990).
209. J. H. Simmons, R. Ochoa, K. D. Simmons, and J. J. Mills, *J. Non-Cryst. Solids* **105**, 313 (1988); J. H. Simmons, R. K. Mohr, and C. J. Montrose, *J. Appl. Phys.* **53** 4075 (1982); R. Yamamoto and A. Onuki, *Enronhvs. Lett.* **40**, 61 (1997).
210. G. Kohring, R. E. Shrock, and P. Wills, *Phys. Rev. Lett.* **57**, 1358 (1986).
211. J. H. Akao, *Phys. Rev. E.* **53**, 6048 (1996); S. R. Shenoy, *Phys. Rev. B* **40**, 5056 (1989); B. Chattopadhyay, M. C. Mahato, and S. R. Shenoy, *Phys. Rev. B* **47**, 15159 (1953).
212. R. P. Feynman, *Phys. Rev.* **90**, 1116 (1953); *ibid.* **91**, 1291 (1953); D. M. Ceperley and E. L. Pollock, *Phys. Rev. B* **39**, 2084 (1989).

213. C. Dasgupta and B. I. Halperin, *Phys. Rev. Lett.* **47**, 1556 (1981); S. R. Shenoy and B. Chattopadhyay, *Phys. Rev. B* **51**, 9129 (1995).
214. H. Kleinert, *Fields in Condensed Matter, Volume II Stress and Defects*, World Scientific, New York, 1989.
215. R. Cordery, *Phys. Rev. Lett.* **47**, 457 (1981); W. Janke and H. Kleinert, *Phys. Lett. A* **128**, 463 (1988).
216. L. R. Corrales and J. C. Wheeler, *J. Phys. Chem.* **96**, 9479 (1992); J. C. Wheeler and P. Pfeuty, *Phys. Rev. Lett.* **46**, 1409 (1981).
217. J. Toner, *Phys. Rev. B* **26**, 462 (1982); D. R. Nelson and J. Toner, *Phys. Rev. B* **24**, 363 (1981).
218. C. Dasgupta, *Phys. Rev. A* **27**, 1262 (1983).
219. N. D. Antunes and L. M. A. Bettencourt, *Phys. Rev. Lett.* **81**, 3083 (1998); M. B Hindmarsh and T. W. B. Kibble, *Rep. Prog. Phys.* **58**, 477 (1995).
220. T. Banks, R. Myerson, and J. Kogut, *Nucl. Phys. B* **129**, 493 (1977); M. Stone and P. R. Thomas, *Phys. Rev. Lett.* **41**, 351 (1978).
221. J. C. Wheeler, S. J. Kennedy, and P. Pfeuty, *Phys. Rev. Lett,* **45**, 1748 (1980); S. J. Kennedy and J. C. Wheeler, *J. Chem. Phys.* **78**, 953 (1983).
222. J. Dudowicz, K. F. Freed, and J. F. Douglas, *J. Chem. Phys.* **113**, 434 (2000).
223. Dudowicz, K. F. Freed, and J. F. Douglas, *J. Chem. Phys.* **119**, 12645 (2003).
224. J. Ruiz-Garcia and S. C. Greer, *J. Mol. Liq.* **71**, 209 (1997).
225. S. C. Greer, *J. Phys. Chem. B* **102**, 5413 (1998); *Adv. Chem. Phys.* **94**, 261 (1996).
226. F. Sciortino, A. Geiger, and H. E. Stanley, *Nature* **354**, 218 (1991); *Phys. Rev. Lett.* **65**, 3452 (1990); *J. Chem. Phys.* **96**, 3857 (1992).
227. J. Stambaugh, K. Van Workum, J. F. Douglas, and W. Losert, *Phys. Rev. E* **72** 031301 (2005).
228. K. Van Workum and J. F. Douglas, *Phys. Rev. E* **71**, 031502 (2005).
229. C. T. Moynihan and J. Schroeder, *J. Non-Cryst. Solids* **160**, 52 (1993).
230. J. P. Garrahan and D. Chandler, *Phys. Rev. Lett.* **89**, 035704 (2002); *Proc. Natl. Acad. Sci. USA* **100**, 9710 (2003).
231. G. H. Fredrickson and H. C. Anderson, *Phys. Rev. Lett.* **53**, 1244 (1984).
232. E. Rössler, *Phys. Rev. Lett.* **65**, 1595 (1990).
233. C. Hansen, F. Stickel, T. Berger, R. Richert, and E. W. Fischer, *J. Chem. Phys.* **107**, 1086 (1997); C. Hansen, F. Stickel, R. Richert, and E. W. Fischer, *J. Chem Phys.* **108**, 6408 (1998).
234. D. Bedrov, G. D. Smith, and J. F. Douglas, *Europhys. Lett.* **59**, 384 (2002); *Polymer* **45**, 3961 (2004).
235. E. Tombari and G. P. Johari, *J. Chem. Phys.* **97**, 6677 (1992).
236. M. Cassettari, G. Salvetti, E. Tombari, S. Veronesi, and G. P. Johari, *J. Non-Cryst. Solids* **172–174**, 554 (1994).
237. R. Casalini, S. Corezzi, D. Fioretto, A. Livi, and P. Rolla, *Cham. Phys. Lett.* **258**, 470 (1996).
238. J. F. Douglas, *Comp. Mat. Sci.* **4**, 292 (1995); J. F. Douglas and D. Leporini, *J. Non-Cryst. Solids* **235–237**, 137 (1998).
239. E. Rössler, *Ber. Bungenges Phys. Chem.* **94**, 392 (1990).
240. D. Prevosto, S. Capaccioli, M. Lucchesi, D. Leporini, and P. Rolla, *J. Phys. Cond. Matt.* **16**, 6597 (2004); D. Prevosto, S. Capaccioli, M. Lucchesi, P. Rolla, and R. Casalini, *Phys. Rev. B* **71**, 136202 (2005).

241. Interaction frustration has also been found[182] to be a control parameter for fragility in a spin model of glass formation.

242. J. De Guzman, *An. Soc. Esp. Fis. Quim.* **11**, 353 (1913); J. M. Burgers, *Introductory Remarks on Recent Investigation Concerning the Structure of Liquids*, Second Report on Viscosity and Plasticity, North Holland, Amsterdam, 1938, p. 34; G. Hägg, *J. Chem. Phys.* **3**, 42 (1935); D. O. Miles and A. S. Hamamoto, *Nature* **119**, 644 (1962); A. A. Vlasov, *Theor. Math. Phys.* **5**, 1228 (1970); M. A. Floriano and C. A. Angell, *J. Chem. Phys.* **91**, 2537 (1989); M. Wilson and P. A. Madden, *Mol. Phys.* **92**, 197 (1970); P. A. Madden and M. Wilson, *J. Phys. Cond. Matt.* **12**, A95 (2000).

243. G. P. Johari, *Phios. Mag.* **86**, 1567 (2006).

244. S. F. Swallen, P. A. Bonvallet, R. J. McMahon, and M. D. Edigers, *Phys. Rev. Lett.* **90**, 015901 (2003).

245. P. G. Debenedetti, *Metastabte Liquids*, Princeton University Press, Princeton, NJ, 1996.

246. J. F. Douglas, J. Dudowicz, and K. F. Freed, *J. Chem. Phys.*, **125**, 144907 (2006).

AUTHOR INDEX

Numbers in parentheses are reference numbers and indicate that the author's work is referred to although his name is not mentioned in the text. Numbers in *italic* show the page on which the complete references are listed.

Advances in Chemical Physics, Volume 137, edited by Stuart A. Rice

SUBJECT INDEX

Advances in Chemical Physics, Volume 137, edited by Stuart A. Rice